RISING MISSION TO MARS

EXTENSIVE COLLECTION OF SPACE SCIENCE RESEARCH PAPERS

A NOVEL APPROACH ON FINDING A LIFE ON MARS ALONG WITH MISSION PROPOSALS FOR FUTURE MARS EXPEDITION

BISWESH DHUNGANA

No part of this publication may be reproduced, distributed, or transmitted in any form or by any means, including photocopying, recording, or other electronic or mechanical methods, or by any information storage and retrieval system without the prior written permission of the publisher, except in the case of very brief quotations embodied in critical reviews and certain other noncommercial uses permitted by copyright law.

Table of Contents

Chapter 1. Investigation of NASA Challenger STS 51-L Accident and Analysis of Engineering Ethics in failure of Shuttle Challenger. 1

Chapter 2. EVA Free Flying Robot (Uni-Robo) Innovation Design 5

Chapter 3. Mars Exploration Rover 12

Chapter 4. The Lunar Exploratory Base Design 24

Chapter 5. Semi-Autonomous Mars Surface Robotic Exploration Mission Proposal (SMSREM) 43

Chapter 6. Challenges and Risks to Human Spaceflight – Risk Management Plan Technical Report 58

Chapter 7. Long Duration Experiment Technical report: Mars Advanced Detector In-Situ-Utilizer 78

Chapter 8. Mars Manned Outpost Mission Proposal ...87

About The Author 112

Other Books By Biswesh Dhungana 114

Chapter 1. Investigation of NASA Challenger STS 51-L Accident and Analysis of Engineering Ethics in Failure of Shuttle Challenger.

Abstract

This case study explores the investigation of what failures were made that led up to the STS-51L "Challenger Disaster" and what better choices could have been made in the process leading up to the launch of the "Space Shuttle". The NASA Challenger STS 51-L was a prototype known to humankind, laying the groundwork for subsequent generations of shuttles to lead in their footsteps and learn never to repeat mistakes. The shuttle was a revolutionary of its kind, with a total of nine missions accomplished, and was the way for subsequent generations of shuttles to follow their footsteps and learn from errors. STS 51-L had its spikes and its puddles, like every other human-crewed spaceflight vehicle. The devastating destruction of Challenger STS 51 – L was solid rocket booster O-rings' failure which caused hot combustion gasses to leak from the side of the boosters, burning through the external fuel tank. However, several factors included the O- ring's failure, including a scarcity of disinformation, Ignorance of Known design faults, desire for Space Race, and proximately the failed responsibility to communicate effectively leading to faulty construction of solid rocket booster that induced its disastrous and devastating demise.

The purpose of the Shuttle Challenger flight was to orbit the second satellite to monitor and relay data for the communications facilities in the United States Although the tragedy happened when the spacecraft was exploded in front of the eyes of the entire company and a significant portion of the American public. The crash took the lives of seven crew members, including a teacher, who had to teach children about their room on their return from the flight. In investigating this tragic incident, it was found that the accident was caused by a malfunction in the O-rings, a failure that was then attributed to a flawed design whose performance could be too easily compromised on the day of launch by factors such as low temperatures. The propeller gasses melted the metals that made up the rocket booster and caused the explosion of the outer tank. There were, however, several other factors that led to the explosion's devastation. Most significantly, coordination and management should have stopped the tragedy. This paper examines the degree to which the ethical ambiguity that pervades the relationship between individual and organizational responsibilities has contributed to the Challenger tragedy in the analysis of these fundamental issues.

The lack of coordination between engineers of various levels of management was one of the unethical practices that eventually lead to the Challenger's destruction. The Roger Commission also found that Marshall managers already knew about the frailty of O-rings as they were tested by the engineers before the launch. Rather than redesigning the join, they neglected and further consulted. Specifically, there was also very poor coordination between Morton Thiokol, the company that produced the Solid Rocket booster and engineers, and NASA management. The outside temperature decreased to 18°F the night before the launch and to 36°F the morning after the launch. Although some engineers thought that even under these conditions, the boosters would still be able to operate efficiently, several were very worried that the temperature would cause a failure. One aspect that eventually led to the catastrophic result is the inability of the Morton Thiokol engineers to effectively express their concerns to the NASA leadership and persuade them to delay the launch of the shuttle.

The ignorance of known design faults was one of the significant causes for the eventual loss of Space Shuttle Challenger. Engineers of NASA had enough time to repair joints for connecting the motor cases with the solid rocket, however, the decision was taken not to redesign the joint, but to regard the problem without further consultation as a reasonable flight risk. Another situation where ignorance was clearly observable was when testing for the O-rings. The O-rings problem was a serious issue in terms of professional ethics because they could not in any way guarantee the desired procedure. After all, the temperatures were not as low as morning, so they were just predicting without any evidence. Because of the disastrous event, despite pressure from the management, the engineers had to change to tackle the problem and had the option to authorize the launch. This clearly demonstrates that priorities were not properly classified based on the level of conflict rather Economic aspect was chosen before protection because delay meant a considerable economic loss. As the night before the launch, Marshall officials warned of the possibility to freeze of O-rings rubber and inflict a catastrophe due to expected cold weather, but company managers and Marshall Project managers declined to call for a launch delay.

In conclusion, the disastrous and devastating demise of NASA Challenger STS 51-L was primarily due to a lack of management and coordination on both Morton Thiokol and NASA as they neglected the effect of the rigid rubber O-ring under cold conditions. However, The Challenger Shuttle laid the groundwork for subsequent generations of shuttles to lead in their footsteps and learn never to repeat mistakes. From its failures, many lessons can be learned to have a firm base in engineering technological expertise, complemented by good management practices.

References

Hastings, D., Launius, R., & Mccurdy, R. (1997). The Challenger Disaster. In Spaceflight and the Myth of Presidential Leadership (pp. 2–6). University of Illinois Press. https://ocw.mit.edu/courses/engineering-systems-division/esd-10-introduction-to- technology-and-policy-fall-2006/readings/challenger.pdf

NASA - STS-51L Mission Profile. (2011). NASA. https://www.nasa.gov/mission_pages/shuttle/shuttlemissions/archives/sts-51L.html

NASA - Space Shuttle Overview: Challenger (OV-099). (2013). Nasa.Gov. https://www.nasa.gov/centers/kennedy/shuttleoperations/orbiters/challenger-info.html

The Crew of the Challenger Shuttle Mission in 1986. (2020). Nasa.Gov. https://history.nasa.gov/Biographies/challenger.html

Washington, D. C. (1986). *Rogers commission report.*

https://spaceflight.nasa.gov/outreach/SignificantIncidents/assets/rogers_commission_repor t.pdf

CHAPTER 2. EVA FREE FLYING ROBOT (UNI-ROBO) INNOVATION DESIGN

Abstract

In spaceflight operations for future EVA flights, human space exploration faces a paradigm change. There are many major obstacles to carrying out crewed space missions while conducting EVA missions. Working in gas-pressurized space suits, however, causes fatigue, unnecessary energy costs and injury. Astronauts should be more autonomous than ever in order to achieve mission success by delayed contact with Earth-based support personnel (Miller et al., 2000). Therefore, an intelligent decision-making assistance robot incorporated with Artificial Intelligence and Machine Learning robot, which supports and supports Crew activities, needs to be designed and built. This study presents the basic design principles from multiple simulation campaigns focusing on the design and scope of the extravehicular activity (EVA) support device used by Free Flying Robot during EVA operations for the performance of intravehicular operations and the autonomous recording of operating noises emitted by individual machines to conduct research investigations. The main feature of this robot is a programming interface, which helps research teams to integrate with the robot with their projects and provides visual navigation markers that are less visual, autonomous docking and charging, energy saving handrails and flexible loads.

Introduction

Many new prospects for exploration are provided by potential human space flights in Earth orbit, to the moon, and to exotic destinations. Astronaut time, however, will always be in short supply, consumables will always be limited, and certain work for astronauts to do manually will not be feasible or efficient. Therefore, enabling crew to be more self-sufficient is a key to ensure mission success. Among the numerous activities astronauts perform, extravehicular activity (EVA) is the among their most challenging and mission critical. NASA aims to develop a long-term strategy for achieving extended human presence into deep-space that include missions stages ranging from cis-lunar space, Near Earth Objects, the Moons of Mars, and ultimately the surface of Mars (NASA, 2016).

The free-flying, named Uni-Robo has the capability, inside the International Space Station, to carry out a variety of tasks. The propulsion system of Robot uses pulses to actively press air into a chamber and then redirects it carefully through one of its six nozzles. Each robot has two drive modules to provide complete holonomic platform control. Uni- Robo packs a tremendous range of sensing, processing, and networking capabilities into a one- foot cube. These capabilities are powered by open-source flight applications running on three interlinked processors that interface with seven embedded systems, six cameras, two propulsion modules, and various other adapters (Smith et al., 2016). The Uni-Robo Robot Program offers mobility and position control, handles different software and hardware modules, supports various operating modes, and accommodate external researcher.

Earth flying robots must overcome high gravity, with limitations that are very different from the ISS flyers. However, multi-rotor drones solve associated uni-robots in a lightweight package in computational efficiency and a disorderly environment. Previous JAXA missions have demonstrated successful remote camera capabilities. CIMON is also another DLR robot with an emphasis on providing astronauts with AI support. The NASA robot is also a complex ISS robot with many similarities to Uni-Robo. The unique feature of Uni-Robo provides Flight Controllers with HD video stream monitoring and a payload capability to conduct a multi-hour survey of specific environmental parameters in the ISS. Guest researchers perform microgravity studies on Uni-Robo with payload software and hardware. The unique feature of Uni-Robo provides Flight Controllers with HD video stream monitoring and a payload capability to conduct a multi-hour survey of specific environmental parameters in the ISS. Guest researchers perform microgravity studies on Uni-Robo with payload software and hardware.

The project Uni-Robot aims to send one dock and four other flyer robots to the ISS. The dock fits with two berths, with power and Ethernet connections linked to the Uni-Robo robot. The dock also has a built-in computer that tracks cabinet halls and Uni-Robo status during the winter season. The docking machine essentially enables robot software updates. E robot. E robot. Each Uni-Robo has 3 ARM processors, which communicate over an Ethernet network, for smartphones class. The "low-level processor" (LLP) operates the posing estimator, motion control circuit and contact with the appropriate hardware such as the IMU and the power management system. The "Mid-Level Processor" oversees computer-intensive viewing and algorithms for mapping. The Navcam and DockCam images are linked to two color pictures and two depth pictures: (Hazcam and PerchCam). The MLP also interacts with the soil and the defect management system. The high level (HLP) processor is devoted to guest science applications developed by external academics and business scientists. The Uni-Robot moves on two propulsion modules, which sandwich the central calculating module with 12 nozzles. Dust forces out the air, which is pressurized by an impeller in each module.

The Uni-Robo is normally fitted with a hanging arm to catch ISS handrails and turn the flying robot into a remote pan-tilt camera. By specifying simple roles for and subsystem, between nodes are often held to a minimum. The photo. Displays distribution on the Uni-Robo processors of the main modules. The robotic program of Uni-Robo is optimized for the high-level structure, support for ROS and performance of C++. Since it is widely used and software package availability (particularly ROS) for this distribution, the LLP and MLP processors run Linux (Ubuntu 16.04 LTS). The movement management module of Uni-Robo operates on the LLP. The close loop control has a jitter smaller than the tolerance agreed by limiting the number of processes operating on this processor. Simulink is used to construct the control subsystem. C code is generated automatically by mapping the Simulink buses into the data structures input/output. The automatically created code is wrapped in ROS nodes, which seamlessly integrate with the rest of the system. The control model parametrization uses the same config files, which are native to the Uni-Robo robot program.

During the test situation, Uni-Robo acts as a microgravity research center. For experimental research and/or presentation in this scenario various sensors and instruments may be connected to the Uni-Robo expansion ports. As appropriate to this situation, Payload developers will upload software or hardware. The ISS crew will mount and assist in the investigation of any payload hardware. During crucial segments of this scenario, the crew can be needed for risk control and/or safety. Accessing, tracking, and supervision of Uni-Robo from the ground with the ISS crew will be provided by a testing facility support engineer at MMOC. The engineer activates Uni-Robo, conducts device reviews, and uploads the schedule for transferring Uni-Robo to the experiment site in order to begin the test scenario. The scheme is being launched and Uni-Robo flies to the testing site autonomously during underground surveillance. The engineer orders Uni-Robo while at the experimental position to hold to a designated handrail and stand up for the next command. The crew then mounts payload equipment and uses the crew control station to set up payload checks. When actually able to do this, the crew triggers and conducts check-out checks to check the capabilities on payload hardware using the Crew Interface. Uni-Robo transmits data during the experiment to the crew and the ground as needed. The engineer also controls Uni-Robo and tests the experimental results. The crew extracts and preserves the payload hardware for potential use after the experiment is over. The engineer then uploads Uni-Robo to the dock and implements the scheme. Uni-Robo reloads the experiment data to the on-board server and downloads them. For more study and analysis, the engineer connects the experimental results.

A strong yet complex hardware platform is controlled by Uni-Robo Robot and Software. By following a dual middleware strategy, we can use ROS onboard while retaining space network constraints. In the future, we foresee the use of DDS as transport layer in the transition into a single ROS2 middleware. The Uni-Robo project's computing architecture is essential to space implementation and maintenance and constitutes a substantial part of the initiative. The embracing of ROS offers a reliable distributed infrastructure as well as access to ROS-including software like Gazebo. Uni-Robo presents a well-built software solution for an integrated robotic platform with ROS. It is an advantageous solution. The prototype Uni-Robo was run for several hours without the lack of a real-time operating system being hindered. The developed software components allowing Researchers to communicate at different levels, depending on their needs, with Uni-Robo robotic software. As the Uni-Robo robots are commissioned on the ISS by 2025, this program provides fresh and exciting research opportunities.

References

Frank, J. D. (2020). Artificial Intelligence: Powering Human Exploration of the Moon and Mars. In *ASCEND 2020* (p. 4164).

Wu, S. C., & Vera, A. H. (2019). Supporting Crew Autonomy in Deep Space Exploration: Preliminary Onboard Capability Requirements and Proposed Research Questions. Technical Report of the Autonomous Crew Operations Technical Interchange Meeting.

Fluckiger, L., Browne, K., Coltin, B., Fusco, J., Morse, T., & Symington, A. (2018). Astrobee robot software: a modern software system for space.

Bualat, M., Barlow, J., Fong, T., Provencher, C., & Smith, T. (2015). Astrobee: Developing a free-flying robot for the international space station. In *AIAA SPACE 2015 Conference and Exposition* (p. 4643).

Yoo, J., Park, I. W., To, V., Lum, J. Q., & Smith, T. (2015, September). Avionics and perchingsystems of free-flying robots for the International Space Station. In *2015 IEEE International Symposium on Systems Engineering (ISSE)* (pp. 198-201). IEEE.

Cauligi, A., Chen, T., Suresh, S. A., Dille, M., Ruiz, R. G., Vargas, A. M., ... & Cutkosky, M.(2020). Design and Development of a Gecko-Adhesive Gripper for the Astrobee Free-Flying Robot. *arXiv preprint arXiv:2009.09151.*

Chapter 3. Mars Exploration Rover
Abstract

This report proposes a remote mission to Mars to set up two mobile science laboratories on the surface of Mars through Mars Exploration Rover to carry out further research on geological investigations, including the characterization of the diversity of rocks and soils that may provide clues to past water activities (Reyl, 2004). The mission to Mars and space architecture is first introduced. Next, the preliminary launch opportunity, transfer of Earth-Mars, capture of Mars, mission orbits, primary navigation schemes are detailed. This paper also addresses the complexities of long-term missions in a spatial environment, opening up the opportunity to seriously consider human enhancement and a genuinely autonomous space science focused on Advanced Artificial Intelligence. Any space program involving long-term human missions will have to cope with severe risks to human health and life (Erik, 2019). As countermeasures currently available are ineffective over the long term, new, more drastic solutions are required such as a human enhancement initiative for potential deep space explorers. Overall, the Mars Exploration Rover provides evidence about past environmental conditions and determines if the environment is life-friendly for future human missions.

Introduction

The Mars Exploration Rover study's main aim is to investigate two landing areas on the Mars surface and determine how well habitable conditions have occurred past aqueous activity at these locations. Two missions are designated as Exploratory-X25 and Exploratory - XX25, launched by Falcon X Æ A-Xii equipped with powerful thrusters from NASA Langley Research Center. The Mars Exploration Mission is divided into ten distinct mission phases, detailed in the Launch and Mission Phases section. This mission uses payload to read the geological record at the landing sites, make compositional and textural rock measurements that will reveal proof of the conditions in which the rocks formed. Science is closely associated with the Mars Exploration Program's objective to determine to which extent Mars provided the appropriate conditions for the production and preservation of prebiotic compounds (Golombek, 2013). This goal is commonly stated as defining Mars's habitability and providing an idea of the roles of tectonics and climate processes in providing life-giving conditions. It is of fundamental importance to provide water and to communicate with crustal materials. Three of the mission goals, therefore, concentrate on investigating past waterproof. Firstly, the goal is to research landing areas that arehighly likely to contain evidence of fluid water action. The next goal is to search for and characterize a diversity of rocks and soils with hints of past water activities and the selection and evaluation of environmental conditions in fluidity and their use. To achieve individual-specific scientific goals, the selection of a suitable landing site is essential.

Mission Objective

The Mars Exploration Rover's objective is to set up a Science laboratory on Mars to investigate the Spatial distribution and composition of the landing sites' underlying minerals, rocks, and soils. By recording the geological and biological background of the landing site, Mars Exploration Rover carries an extremely capable payload to the surface that will enable sample collection, effectively assembling the "field notes" for the sample package. Important newscientific discoveries about Mars, past and present, will be created by the exploration process needed to record field background for a returnable sample collection. These results will advance planetary science, but they also provide the crucial new groundwork for future human exploration of Mars' surface. This mission will follow an unguided, ballistic descent. A heat shield and parachute will be used to slow down its descent through the Martian atmosphere, fire back rockets to reduce its landing speed, and then deploy airbags to coast its impact with the surface (NASA, 2020). The landers retract the airbags and deploy the lander petals after the airbag mounts to a stop. Then the Rover deploys its solar panels, completing the mission Entry, Descent, and Landing (EDL) process (NASA, 2020). On a postgraduate trajectory, Rover will reach Mars and will land early in the Martian afternoon. Earth will be viewed during EDL to allow signal tones to be transmitted, coded for crucial steps to be taken during the EDL timeline. In addition to the direct-to-Earth (DTE) X-band signals, the spacecraft Mars Global Surveyor (MGS) will be called into operation through a UHF connection to relay descent information (Portock, 2002). One of the most critical landing safety issues is high winds, as high winds worsen landing risks due to rocks and slopes. Recently, due to concerns about high horizontal winds, two exciting sites in the Valles Marineris canyon have been eliminated (Golombek et al., 2003) As a consequence, it has become a priority for this project to identify wind patterns in landing sites, even to the extent of relaxing some of the other engineering constraints. The two power and telecommunications' capacity will be decreased by the end of the Surface Operations process as the Earth, and Sun gets farther from Mars, the dust falls on the plates, the batteries lose their capacity, and Sun travels farther north past the latitude of the landing site. The Rover's ability to drive significant distances and the time the Rover can keep awake will decrease every day. Finally, often after Sol 91, the Rover is projected not to be in a position to store enough thermal or battery power to prevent overnight temperatures of its components from dropping below permissible flight levels (Vasavada et al., 2014). The cruise stage provides most of the spacecraft's specific subsystem functionality during the interplanetary transition to Mars (Guinn, 2008). However, the cruise process is operated by the flight computer inside the rover. The propulsion system consists of a 3:1 blowdown-range monopropellant hydrazine system. It has two spherical titanium propellant tanks and two diametric thruster clusters with four 4.4-N thrusters, each of which is opposed. The array is divided into five parts, three of which can be triggered or disconnected via ground control, if necessary. There are two X-band antennas in the crossing phase system: a low gain antenna (LGA) and a medium gain antenna (MGA) (James, 2005). Both the standard solar array and the LGA and MGA point towards the Z-axis spacecraft aligned with the spacecraft's spin axis. The Small Deep Spatial Transponder (SDST) facilitates the two-way Doppler, ranging system, the removal and identification of the command signal, telemetry, modulation coding, and differential one-way tone generation

(DOR) at ±19 million is a core component of the X-band communication system.

Mission Phases

Mission	Exploratory-X25 Start (EST)	Exploratory-XX25 Start (EST)
Launch	February 5, 2024	March 1, 2024
Cruise	February 6, 2024	March 2, 2024
Approach	October 6, 2024	November 12, 2024
EDL	December 5, 2024	January 12, 2025
Post-Landing	January 25, 2025	February 28, 2025
Surface Operations	January 25, 2025	February 28, 2025
End of Mission	May 2, 2025	June 1, 2025

Launch and Mission Duration

The launch commences when the spacecraft moves onto the launch pad to internal power and ends when the spacecraft is deemed secure, safe and the startup telemetry is played back. The key activities during the mission phase are the launch vehicle lift and ascending phase, entry into a circular parking orbit, coastal cycle followed by the additional launch vehicle burns required to fill the spacecraft with an interplanetary path to Mars, spacecraft separation from the launch vehicle, initial acquisition by the DSN and testing of the spacecraft (NASA, 2021). The double-launch time architecture was designed to meet a range of competing criteria and limitations for the Exploratory - X25 and Exploratory -XX25 missions. This involves trying to remain within the Delta II launch vehicles' performance limits while preserving the view of the Earth over EDL, adapting to the various launch pad restrictions, and ensuring a reasonable chance that both spacecraft can launch during their respective launch times. In developing the overall launch period strategy, other design priorities were also important, such as using a regressing arrival for each MER mission to be planning (Marvin, 2016). Two different starting opportunities typically occur each day to launch a rocket and inject its payload into the correct Earth escape trajectory for any given launch azimuth. The primary distinction between these opportunities is when a when must coast in a parking orbit until it reaches the appropriate position for carrying out Earth injection burning, typically known as long-coast and short coastline opportunities (Salotti, 2004) The short-coast opportunities culminated in much longer solar eclipse times after the separation of the starship from the Mars Exploratory missions' launch vehicle. Two spacecraft from the Mars Exploration Rover will be deployed independently on routes to Mars from late Jun to mid-July 2024. The starship concept is based on Mars Pathfinder's cruise, entry, descent, and landing configuration. They hit Mars on 6 February and 27 February 2025, directly penetrating their interplanetary paths into the Martian atmosphere and land on the surface at two different landing points. Everyone can then throw its landing shell and reconfigure it into a revolving science laboratory that will investigate the landing sites' geology using a range of instruments and equipment, both remote and in situ (Sherwood al., 2002). The vehicles will stay on the surface of Mars until May 2025 for over three months each. Their missions will continue into the winter of 2025, depending on the Martian climate and the systems' survival.

Mission Users

A remote-controlled Mars exploratory Rover offers the potential for enormous quantities of scientific data to be collected; Mars is an ambitious biological goal. However, the increased risk cannot be explained by discovery alone. An economically viable mission will fuel growth in related high-tech industries by introducing a human mission research and development program. Owing to its relative proximity to the Earth and its analogous surface conditions, Mars offers a viable destination. There is no shattering gravity or toxic atmosphere, which renders it impossible for people to explore other solar system bodies. Moreover, Mars is both a scientifically fascinating and desirable goal to the general public. Mars has fascinated people and continues to do this since early Greek mythology, as is evidenced in the recent movies and books on pop culture. Four decades of robotic study by the United States and Russia have made us aware that Mars, at least in the past, was not exactly a world unlike ours (Ehlmann et al., 2005). The robotics science program has focused on the interpretation of the geological and climatic history of Mars. For living beings, liquid water is essential, and Earth's water history determines paleo habitability. In recent decades, researchers on Earth have found microbes emerging in very inhospitable conditions in extreme environments, almost everywhere where liquid water is present. It suggests that life on Mars was possible, following what we now know about Mars' abundance of water. Furthermore, the study of Mars' habitability enables us to answer basic questions and better understand the forces that form our world's life history. Mars, then, is the ideal NASA scanning destination. Orbital and landing craft for human lunar missions have been developed; however, interplanetary propelling is a new undertaking for a human-machine (Ludwinski, 2012). Propulsion is vital to any proposed Mars mission's success and decreases astronauts' exposure to radiation and microgravity by reducing transit time. Fuel is essential for speeding up the rocket, but the spacecraft mass and the starting costs are also increased by additional fuel. These techniques are well understood to improve further and use liquid, effective, and nuclear propulsion technology. (Thirsk et al., 2009) Both cost and transit time, an optimal combination must be chosen. The technologies used to reduce fuel costs and speed can be used, such as the aerobraking architecture, where cargo is first dispatched, and astronauts are then dispatched into faster spaceships. Pilots have also investigated the use of Mars carbon dioxide as a return fuel. The findings are substantial analytical results. However, the mission is much more persuasive to US policymakers because it will generate innovations, strengthen international ties and empower scientists and technicians of the next generation (NASA, 2020). This will take place at a reasonably low rate. We cannot wholly determine any issues until we start a mission (Norris, 2005). There will be some problems. The technology aspects can only be prepared and addressed at some stage in advance – the growth and development of technology to spend capital. The development process will start with the technology that we will find on Mars by setting a human mission's objective. This development-concept focused on priorities dates back to the Apollo period in a previous study of the viability of a human mission to the Moon by the Space Studies Commission (Hellweg, 2007). They concluded that without a specific human exploration of solar system systems in technological planning, it would eventually be omitted as the steps required to protect these bodies were not considered (Ehlmann et al., 2005). Therefore, it would be impossible that the technology needed to make the journey to Mars would be established unless people were explicitly defined as an integral part of NASA's Mars exploration.

Mission Constraints

The terrain is the largest single source of danger for Mars rovers. Curiosity has experienced an unpredictably high wheel damage rate, particularly on Sols 450-515. This presents a challenge, particularly for the operation of an MSL-class rover, since, on the one hand, it must be very risk-averse in order not to lose the precious rover, while, on the other hand, because of its difficulty, it already needs a considerable amount of labor (Lakdawalla, 2014). The rover should have vision-based terrain classification ability to accurately recognize terrain types and rock characteristics, such as pointiness and embeddedness, and risk-aware route planning ability to recommend safe paths in terrain types, hills, and positive and negative obstacles, to solve this problem. To capture human experts' expertise and incrementally boost output as a mission accumulates data, the Mars Exploratory Rover is coded with a machine learning approach. First, a training data set is created by human experts, consisting of images with labels defining terrain types. When planning a road, different safety standards must be respected. The rover must avoid any potential risks to its wheels, mainly angular embedded rocks, from being driven over. The rover can drive over enormous obstacles, but rocks more extensive than 30cm are avoided in the interests of rover safety, and rocks more extensive than 10-15cm are typically also avoided. The amount of turning during a drive is reduced to protect the wheels and the rover is pushed either forwards or backward to keep the number of turns minimal. The rover planners often consider the terrain's slope and keep the current navigation imagery within the distance limits. Rover operations are necessarily a labor-intensive process to satisfy all safety standards, and for preparing a route to ground, often 30 m – 70 m long, it usually requires 3 people to work 8-10 hours. et minimizing costs, reducing the cognitive load of rover operators, avoiding human mistakes, and, most importantly, dramatically reducing the risk of rover failure, this new risk consciousness-based rover operation would bring benefits to Mars rover missions (Crisp et al., 2003). The Rober-Risk identification method autonomously detects dangerous ground embedded in the Mars Rober Exploratory, recommends secure paths that avoid risks. A vision-based field classifier with a Random Forest machine learning algorithm was developed on the potential for risk assessment. The classifier was trained by an on-board image training collection composed of an expert in the human subject matter. Each pixel of an image is categorized in five terrain types: sand, loose rock, base, embedded pointy rock, and integrated round rock ("NASA Technical Reports Server (NTRS)," 2021)

The Mars Exploration Rover Mission will be the first extended exploration with a mobile robotic science laboratory of another planet's surface and will provide the basis for the next generation of surface science missions. The scalable computer architecture enables rovers to perform concurrent payload and engineering operations, effectively utilizing the time when the rovers are ready every day. Each rover has a collection of tools to perform a field geological investigation of each landing site and define a range of scientific goals to understand water's role in the past environment on Mars. Accelerometer and gyro readings taken during the atmospheric input of a spaceship and the sky's remote sensing after landing will be used for measurements related to the Mars atmospheric science. Engineered instruments are used to supply additional science research opportunities such as regolith excavation, inspection, rock movements, and dust accumulation with wheels, cameras, robotic arms, and solar surfaces. The team has the equipment and training needed to simultaneously procure, evaluate, and schedule daily operations for two rover vehicles. This data set will permit the scientific community to continue to examine in more detail and generate new observations and findings long after the mission'sconclusion.

References

Crisp, J. A., Adler, M., Matijevic, J. R., Squyres, S. W., Arvidson, R. E., & Kass, D. M. (2003).Mars Exploration Rover mission. *Journal of Geophysical Research: Planets, 108*(E12). https://doi.org/10.1029/2002je002038

Ehlmann, B. L., Chowdhury, J., Marzullo, T. C., Eric Collins, R., Litzenberger, J., Ibsen, S., ...Douglas Grant, F. (2005). Humans to Mars: A feasibility and cost-benefit analysis. *Acta Astronautica, 56*(9-12), 851–858. https://doi.org/10.1016/j.actaastro.2005.01.010

Golombek, M. P., Grant, J. A., Parker, T. J., Kass, D. M., Crisp, J. A., Squyres, S. W., ... Fergason, R. L. (2003). Selection of the Mars Exploration Rover landing sites. *Journal ofGeophysical Research: Planets, 108*(E12). https://doi.org/10.1029/2003je002074

Guinn, J. R., Garcia, M. D., & Talley, K. (2008). Mission design of the Phoenix Mars Scout mission. *Journal of Geophysical Research, 113*. https://doi.org/10.1029/2007je003038

Hellweg, C.E., Baumstark-Khan, C. Getting ready for the manned mission to Mars: the astronauts' risk from space radiation. *Naturwissenschaften* **94**, 517–526 (2007). https://doi.org/10.1007/s00114-006-0204-0

J. Salotti, R. Heidmann and E. Suhir, "Crew size impact on the design, risks and cost of a humanmission to mars," *2014 IEEE Aerospace Conference*, Big Sky, MT, 2014, pp. 1-9, doi: 10.1109/AERO.2014.6836241.

Lakdawalla, E. (2014, August 19). Curiosity wheel damage: The problem and solutions. Retrieved January 25, 2021, from https://www.planetary.org/articles/08190630-curiosity-wheel-damage.

Ludwinski, R., Roncoli, R., Ludwinski, J., Thunnissen, D., & Desai, P. (2012, June 25). MissionDesign Overview for the Mars Exploration Rover Mission. Retrieved January 21, 2021, from https://arc.aiaa.org/doi/10.2514/6.2002-4823.

NASA Technical Reports Server (NTRS) | USC Libraries. (2021). Retrieved January 25, 2021,from Usc.edu website: https://libraries.usc.edu/databases/nasa-technical-reports-server- ntrs.

NASA Technical Reports Server (NTRS). (2021). Retrieved January 25, 2021, from Nasa.govwebsite: https://ntrs.nasa.gov/citations/20140009943

Portock, B. (2003, January 25). Mars Exploration Rover Orbit Determination Using VLBI Data |Guidance, Navigation, and Control and Co-located Conferences. (2012). from Aiaa.org website: https://arc.aiaa.org/doi/10.2514/6.2002-4814

Sherwood, B., Smith, D., Greeley, R., Whittaker, W., Woodcock, G., Barton, G., Pearson, D., &Siegfried, W. (2002). Mars Sample Return: architecture and mission design. *Proceedings,IEEE Aerospace Conference, 2*, 2-536 vol.2.

Thirsk, R., Kuipers, A., Mukai, C., & Williams, D. (2009). The space-flight environment: the International Space Station and beyond. *Canadian Medical Association Journal, 180*(12),1216–1220. https://doi.org/10.1503/cmaj.081125

Vasavada, A. R., Grotzinger, J. P., Arvidson, R. E., Calef, F. J., Crisp, J. A., Gupta, S., ...Yingst, R. A. (2014). Overview of the Mars Science Laboratory mission: BradburyLanding to Yellowknife Bay and beyond. *Journal of Geophysical Research: Planets, 119*(6), 1134–1161. https://doi.org/10.1002/2014je004622

CHAPTER 4. THE LUNAR EXPLORATORY BASE DESIGN

Abstract

This report proposes the design aspects of the lunar base, structural loads, and other criteria for human habitation design, with possible materials and structural concepts for second-generation construction on the Moon and an emphasis on space exploration (Ruess, Schaenzlin, & Benaroya, 2006). Further, this paper will consider certain factors like radiation protection and meteorite safety. Additional attention is laid on problems related to mechanical construction and regolith mechanics. A lunar baseline proposal is presented in the paper to analyze potential scientific and design estimates and estimates of costs and construction timelines. This paper also demonstrates how to model lunar landing terrain with typical statistical features, including craters and boulders, to build a basal lunar surface using Brown Fractal Motion (Yibing, Ronghui, Linhui, Lie, & Mingheng, 2014). The next step is to add craters and boulders to the known diameter algorithm and Random-create Diameter Algorithm. Overall, an overview of potential structural concepts is provided, including descriptions of the current lunar environment that engineers must design against—following a preliminary study on constructing the basic lunar structure for the living room. Finally, the paper concludes with an introduction to the structural challenges faced bydesign.

Since long before the dawn of the space age, concepts have been proposed for lunar base structures (Benaroya, 1994). Since the Apollo program days, significant studies have been carried out when it seemed likely that the Moon would become a second home for humans (Benaroya, n.d.). In future, people hope planetary vehicles will be run freely in more sophisticated and stiff environments to study planet exploration activities further. It can thus reveal the World, its creation, and its changes. It is also concerned with the use of lunar regolith and other problems in science engineering. (Yibing, Ronghui, Linhui, Lie, & Mingheng, 2014). The focus here is on human housing structures, a technically demanding fraction of the overall number of systems that the lunar facility is likely to contain. Any proposed lunar base system is checked to see if it meets clear and criteria.

All designs on the lunar surface, different from those for terrestrial forms, must meet various constraints. Several generic structural types are proposed for lunar base structures. That includes concrete structures, metal frame structures, pneumatic and hybrid structures. There are also options for using natural features for sub-surface architectures, such as lava tubes. In theory, each of these approaches can encounter different and different constraints (Benaroya & Bernold, 2008). The Moon base considerations are more practical today: astronomy, mining, or tourism as potential arguments for human existence on the only natural satellite on Earth. Another crucial aspect of a lunar structural design is assessing the total life cycle: the retirement and disposal of a system from conception or recycling the system and its components. Due to the lunar climate design and the inability to assess the system before it is designed and used, several factors influencing system existence cannot be expected (Benaroya, n.d.).

This paper is aimed at addressing technical problems and presenting historical contexts. There are crucial problems here, such as funding the Moon's return, human understanding physiological and many others. The emphasis is on delivering a brief description of the structural and structural engineering challenges for human housing on the Moon and the construction problems. The development of a comprehensive design and prototyping are essential components in a design process (Benaroya & Bernold, 2008). Another crucial aspect of the lunar structural invention is evaluating the full life cycle, i.e., the system's retirement and disposal from conception or recycling the system and its components ("Space Place | ASCE," 2014). Owing to the nature of the lunar environment and the inability to test the device in practical order before construction and use, many factors affecting system life cannot be anticipated. Simultaneous engineering appears to be part of LMS science, design, and construction (Benaroya & Bernold, 2008).

Lunar Base Development Phases

In around three stages, the construction of a lunar base can be planned. These phases will not necessarily be discreet, as all three may coincide at some future point in time. The subsequent steps will depend on developing a more significant body of lunar knowledge and expertise.

Phase 1

The supply marks the first phase of the lunar base construction to the lunar surface of pre- constructed and pre-fitted modules. A minimum of construction operation is expected, such as site levelling or construction of foundations. The only component assembly likely to occur is the attachment of modules together, airlocks' connection to modules, and utility links. It is alsoexpected to be essential to deploy and change support or levelling systems (Smith, 1993). By non- destructive testing of products and assemblies, it can keep material defects to a minimum level. The modules should be packaged to minimize the chances of lunar surfaces being minimized during startup, space travel, and landing. The probability of the material used in step 1 is below fracture or fatigue failure. It is also probable that, depending on Moon and space exploration missions' nature and length, lunar installations' life cycle will be relatively limited.

Phase 2

The second phase of lunar base development will be assembling components fabricated on the Earth and transported to the Moon. One structure concept describes an inflatable sphere several meters in diameter with an internal structure erected from columns, beams and bracing. Some assembly is required. As in the first phase, the components can be carefully fabricated and analyzed before shipped (Benaroya, 2018). Connections and fasteners should be simple and easy to install. Outfitting the structure with interior equipment must be carefully planned, and it must follow the plan to avoid overstressing components or joints.

It will be necessary for phase 2 infrastructure to run, manage, repair, repair processes and materials. To ensure compatible techniques and strategies, the materials and procedures for this step should, where possible, be evolved from those for Phase 1.

Phase 3

The third step of creating the lunar base would include constructing more and more large structures made up of indigenous materials. The components would have more inferior quality and properties, and it more challenging to test the component pieces. Thus, design parts are possibly considerably larger to mitigate potentially lower material durability. Essential components such as screening materials and internal equipment are still needed from Earth. Yet, human beings are considered relatively self-sufficient at this point in the creation of a lunar base. The large structures will contain conditions required for agriculture and operate at atmospheric pressures well below the average Earth level pressure (Smith, 1993). The advantage of using the materials derived in the building from lunar materials is that the lunar atmosphere's conditions are reasonably compatible in a few decades. It can weaken these materials to a minimal degree. As with the other two phases of lunar base construction, they establish operational, maintenance and repair plans, and materials will be a necessity in this process (Rochette, Gattacceca, Ivanov, Nazarov, & Bezaeva, 2010).

Timeline of Mission Events

On March 10, 2022, Lunar Explorer will blast off to the Moon aboard a three-stage StarshipX III solid-fuel aboard NASA

Langley Research Center. It will be successful on its way to the Moon for a three-year polar orbit, a primary mission dedicated to the global mapping of lunar resources, gravity and magnetic fields, and even outgassing events. The StarshipX III will be put in the Lunar Explorer payload approximately 15 minutes after launch into a parking orbit 200 miles above the Earth. The Explorer's Trans Lunar Injection (TLI) stage will complete a 64-second burn following a 45-minute coastline in the parking orbit, release the spacecraft from Earth orbit and set it on track to the Moon, a 105-hour coast. When the spaceship changes to 56 minutes, 30 seconds after the shutdown, the official mission timeline will begin (Benaroya & Ettouney, 1992). Mission controllers will be deployed within the three extendable masts or spacecraft booms shortly after the vehicle is turned on. The spacecraft's five instruments will activate the gamma spectrometer, alpha particle spectrometer, neutron spectrometer, magnetometer, and electron reflectometer. The Lunar Explorer will be successfully captured into lunar orbit on January 13, 2022 and will begin its mission to map the Moon globally a few days later.

In constructing a Lunar Base, it will address the importance of studying the mechanics of materials. The entire picture consists of materials alone, but materials used for building lunar-basedinfrastructure must work as expected in the Luna environment with fracture and fatigue properties. There are numerous sources of extensive fracture studies and technical literature solutions, and therefore it gave no specifics here (Ganapathi, 1993). Similarly, materials (excluding derived lunar materials) and design techniques exist for fatigue. For problems in which composite materials are used, it can quickly obtain unique methods. Both the classical fracture and the ductile fracture may be actual and, in some cases, the same material, as the lunar environment varies considerably. It cannot be overemphasized the need for thorough design study and comprehensive testing before pursuing 'finally' material applications in the lunar base infrastructure.

Location

To the success of the project, the position of the base is of great importance. The Lunar North Pole's basis position is chosen for the principal living module on a lunar latitude plateau 1/4 89.86. The lunar longitude 109.6, situated at the lunar north pole, will be selected as the colony's location to have the best chance of survival. The telescope lies inside a crater with 1/489,9 lunar latitude coordinates, 1/4 147,34 lunar duration. This location is chosen because it must seek more permanent foundations at a micro-precision scale since the precise position is far more influential in these bases' success. The future work will be followed by a detailed study of the LB10-H exterior and interior design. Concerning base sites, a cross-calibration of data between LRO and GSRR includes a review of the collection. Further scientific targets for base selection will be included, in particular, with data on lunar ice ("Moon's South Pole in NASA's Landing Sites," 2019).

Environment

The problem of designing a structure to build on the lunar surface is a difficult one, discussed here in a necessarily cursory way. Many issues are not discussed but will need to be tackled eventually. Some essential topics not discussed here but necessary in a detailed study, including Relationships between severe lunar temperature cycles and structural and material fatigue, a problem for exposed structures; Anaerobic decomposition of exposed steels and other high vacuum effects on brass, alloys, and advanced materials; Reliability and risk must be significant components of lunar structures, just as they are of substantial Earth structures (Benaroya, 1994), Consideration of new failure modes such as those due to high-velocity micrometeorite impacts. The development of a comprehensive concept and prototyping is an integral component of the design process. Such constructions and practical experiments cannot be carried out on Earth or even in space in a lunar environment setting. For example, at this time, the effect of 1/6 g suspended lunar regolith lunar soil fines on lunar machines cannot be tested experimentally. Experience in Apollo can be extrapolated, but only to the limits beyond which new knowledge is required (Maiwald et al., 2020). The assessment of the overall life cycle, from construction to retirements, or the recycling and re-use of structures and their components, is another prominent feature of a lunar structural design. Due to the lunar climate's existence and the inability to test the system in realistic order before building and using it, several factors influencing system life cannot be forecast.

In the end, simultaneous engineering tends to be a part of LMS research, design and construction. Concurrent engineering considers device design, fabrication, and installation, bringing significant components to a stage as early as possible to expect future issues. This description has an additional dimension here. Given the intense nature of the structure's environment, competitions must require versatility in design and construction. Parallelism should be retained in the design area so that alternatives exist at each juncture, which would enable the building to proceed even in the face of totally unforeseen difficulties. This aspect requires further attention and a thorough analysis of its consequences (Maiwald et al., 2020). There has already been a discussion of lunar architecture codes (Benaroya and Ettouney; 1992). In recent years, the use of gallium-arsenide has shifted away from electronic silicone and germanium-based parts. Reducing current and demand for voltage and miniaturizing the electronic components and machinery will improve radiation. It can simulate cosmic radiation effects using radiation transportation codes since it cannot achieve this in the laboratory. LAHET [34], developed by the Los Alamos National Laboratory, is one of the most potent codes found (Duke, Mendell, & Roberts, 2015).

Vacuum

A difficult vacuum surrounds the Moon. This prohibits the use of certain materials that, under certain conditions, cannot be chemically or molecularly stable. Another significant issue to take into Consideration is vacuum blasting. When the explosive is fired into a cavity, it becomes a gas with pressure, often reaching 100,000 terrestrial atmospheres. It's hard to predict how this will affect the region around the Moon explosion and the effects of the blast (Benaroya, Bernold, & Chua, Koon Meng, 2021). A layer of fine particles has the lunar surface, which is quickly disruptedand suspended. These particles are connected to all characters and pose severe challenges for building equipment, airlocks and exposed surfaces. Lunar dust is composed and tends to be charged with pulverized regolith. The charge may be due to the broken crystalline structure or surface-like costs such as charged solar-wind particles attached to the dust-parts.

Operational Concept

The operational Concept for lunar base structures has been proposed since long before the dawn of the space-age. As these are likely to form the pool of potential lunar baseline designs, it will only generate this report during the last quarter of a century. It is a small but technological portion of the total number of structures that are likely to comprise the lunar facility that is the priority. The buildings consist of processing and mining areas, crew members' living areas, laboratories, and greenhouses. For the crew to house all medical equipment and supplies, there is also a designated emergency room. It has computers that can help the medical doctor diagnose the illness, manage it, or treat the patient himself (Pathways to Exploration, 2014). The external defibrillator, which is operated by electrical shocks to restart the heart of the patient, is an example of such an instrument, but may also tell the operator if the patient needs defibrillation, whether he or she has the appropriate course of action to administer the therapy (Erickson, 2011a). The criterion for any proposed lunar base structure is how it fits the basic and special requirements outlined in the last section. Many conditions must be satisfied on the lunar surface, distinct from those of the terrestrial structures. Various structural methods have been proposed for lunar base structures. That includes concrete structures, metal frame structures, pneumatic structures and hybrid structures. Besides, subsurface construction options exist, and natural features such as lava tubes are created and used. Theoretically, each solution should satisfy the various, but multiple, constraints (Pathways to Exploration, 2014). Factors that need to be addressed in the design of lunar structures include vacuum, low severity, significant temperature variation, micrometeorite bombardment and radiation from charged particles. As structures with regolith, vacuum, and low gravity can be reduced to cover the last three conditions, the structural design seems to be much more critical. The configuration of long-term structures is influenced by low gravity, while a vacuum affects the design of pressure structures such as habitats very much (Colwell, Batiste, Horányi, Robertson, & Sture, 2007).

One of the problems for the society of extraterrestrial systems is building. Lunar techniques of construction vary from those of the Earth, e.g., the building team will work in pressurization suits. The movement will be dominated by a sixth g, solar and cosmic radiation not covered by the Earth type (Sambiasi, 2019). Various building methods for structure groups and their respective materials are measured. It shall address a habitable moon structure using concrete modules, structural and architectural designs, and manufacturing plants and building methods. The module can be broken down into panels and frames. The Concept and construction of a lunar outpost installation is a qualitative analysis carried out. It would use this facility to construct buildings in one piece that would be too massive to transport the Moon. It will also use the assembly facilities during the lunar outpost running to facilitate operations and maintenance ("Lunar Bases and Space Activities of the 21st

Century," 2019). A variety of trade studies on constructing such an assembly plant have been suggested.

Uses of Inflatables in Lunar Base

A pillow-shaped structure was suggested as a possible concept for a permanent lunar base (Vanderbilt et al. 1988). Using fibre composites, the proposed foundation consists of cylindrical, inflatable, pressurized tensile structures. An abundance of regolith offers a protective cover that accommodates sunlight. Nowak et al. (1990) continue these inflatable design studies concerning the fundamental problem and additional reliability problems. This Concept differs significantly from many other inflatable ideas in demonstrating an alternative to spheroidal inflatables and optimizing housing sizes. A broad lunar base was proposed to simplify and accelerate the process and reduce costs, with inflatable structural ideas (Broad 1989). it suggested a pressurized membrane structure for a continuous lunar base (Chow and Lin 1988,1989). It is made of structural foam with a double membrane on the skin. A pressurized torus-shaped substructure will support the edge support. A regolith excess will provide security. In a short time, the construction process involves the floor's form and the uninflated structure's distribution. The structural foam is pressurized into the internal compartment. The base of this inflated structure is filled with compact flooring that ensures stability and a flat surface within

Construction in New Environment

The installation will also support the lunar outpost's service and maintenance throughout the useful life (Sherwood 1990). A variety of commercial studies on the building of such an assembly plant are planned. At least part of the lunar base development would rest on the Army Engineering Corps' expertise. Preparatory work is now in progress, and it addressed the problems (Simmerer 1988) (Sargent, Hampson, 1996). In the context of an inhabited structure on the Moon, concrete modules address structural and architectural designs, production plants, and construction (Nambaet al., 1988). The module can be split up into panels and frames. A qualitative survey of a lunar outpost assembly system (Drake, Richter, 1990) was conducted. Such an arrangement will be used for constructing too big a structure in one piece for transport to the Moon. Concrete modules are used in structural and architectural designs, manufacturing facilities, and building methods in the sense of an inhabited Moon system (Namba et al., 1988). The module can be divided into frames and tables. It qualitatively investigated the lunar outpost assembly system's architectural characteristics (Drake, Richter, 1990). This is an arrangement to create a device that is too massive for transport to the Moon in one piece.

Transportation

The key aims of road base and road surface are the spatial loads to the full permissible bearing capacity, to provide a stable resistance to traction for required wheel rim pulls – force on the wheel's edge to enable movement! and for abrasion resistance and breaking points. A considerable weight, gravitational force and abundant power source depend on soil-based devices to achieve these goals. While it converted the power source from a diesel engine to electricity, its size would make it prohibitive to install on the Moon. Furthermore, (Bernold 1994b) showed that the compaction of lunar soil needed for a secure road base causes specific issues. Preliminary research indicates that soil particles' normal distribution employing traditional static and vibratory compaction methods will not achieve the required density and power. A considerable weight, weight and energy source relies on soil-based devices to meet these objectives. The power source would be prohibitive on the Moon as it converted it from a diesel engine to electricity. Moreover (Bernold, 1994b), lunar soil's compaction essential to a stable road base has been shown to cause unique problems. Preliminary research indicates that soil parts' normal distribution would not achieve the necessary density and power via the conventional methods of static and vibrational compaction. It is certainly advantageous to use multipurpose equipment but slow the process on the other side. If you want to rely on "proven earth' technologies,' you will have to overcome significant disadvantages. The biggest challenge to the efficient operation of roadbuilding devices is the substantial reduction in gravitational accelerations.

Robotic Operation of Construction Machinery

A significant problem with robotic equipment control is that the signals need to go through the transmission lines of vacuum, air, or fibre. It is generally called latency as the signal travels through the network. (Nelson et al. 1998) Late-service report: 'Teleoperation is commonly used in terrestrial mining equipment remote control. Earth's teleoperating mining equipment is tempting but needs a time delay of 4 to 10 seconds for transmission loops. A human operator can manage time delays of about 1 second in simple teleoperation applications.' Several control schemes have been developed that help to ease the problems caused by excessive signal delays. They range from teleautonomous to predictive Teleoperation to semi-autonomous operation

(Conway et al. 1990).In this context, the research effort is to equip mobile computerized equipment with robustness and intelligence to adapt to the environment's complexities. To achieve this, it is necessary "to create complete agents operating with real-time sensors in complex environments. External models full of external world representations are not sufficient for agents to behave competently and be challenging to obtain. "(Brooks 1986, 1990). Since building work requires a dynamic environment to move inside and deal with while using messy materials that need to be attached, lay on top. A distributed intelligence found on the site equipment, sensors and network communication could be an exceptionally skilled approach for creating a semi-autonomous equipment fleet.

Psychological needs and Impact of Lunar Mission

Behavioural health problems are natural psychological and interpersonal responses to outside- ground situations in the sense of space travel. Psychiatric disorders, however, represent an abnormal reaction. The existence of psychiatric difficulty may lead to a specific person's genetic, constitutional and emerging vulnerabilities. But operative task forces can also play and must be taken into account, including organizational stressors and psycho-social variables including crew conflict, morale, leadership, and cultural and language differences. It is necessary to state that remote and contiguous habitats may also be growth-enhancing and salutogenic before considering these psychiatric issues ("Astronaut selection and medical requirements," 2009). For instance, people may experience more extraordinary power, perseverance, freedom, autonomy, naivety, comradeship and even lower stress and depression in the polar environments or spaces. Any astronauts or cosmonauts in space have recorded transcendental encounters, theological inspiration or a greater sense of human unity by gazing at the World below and beyond. Thus, certain people will benefit from participation in space missions and related environments ("Astronaut selection and medical requirements," 2009). Crew members and staff with mission control must engage, often together, in preparatory training. Hence, the two groups rely mutually on one another in performing mission operations and may also participate in maladaptive communications patterns. Potential training topics for briefs include crew stress, teamwork and missionary relationships, influence on interpersonal interactions of cultural differences, and the proper use of different leaders. Crew members and staff with mission control have to engage, often together, in preparatory training. Hence, the two groups rely mutually on one another in performing mission operations and may also participate in maladaptive communications patterns. Potential training topics for briefs include crew stress, teamwork and missionary relationships, influence on interpersonal interactions of cultural differences, and the proper use of different leaders.

Personnel

The team will be made up of 17 members, making room for the additional crew if the mission is extended. Members must be from the United States, and most hold US Citizenship. There will be no defined gender ratios for the crew to allow the most eligible individuals to be chosen. The selection process would be based on the selection process will be based on criteria set by NASA, which includes having at least two years of related professional experience obtained after degree completion or at least 1,000 hours of pilot-in-command time on jet aircraft and to be able to pass the NASA long-duration flight astronaut physical. There will be three steps for selection. The first step will be filling the application with information about states and contracts. The process will consist of four rounds. Round one will require the applicant to apply and a short video which states why they should be selected to go to the Moon. Applicants must be at least 18 years of age so that they can enter legal contracts (Musson, Sandal, Gro. M, & Helmreich, 2018). There is no age requirement; if you follow health requirements, anyone can apply. Round two will consist of medical and psychiatric assessments and individual interviews.

Officials are looking for endurance, adaptability, curiosity, optimism, and creative thinking in the talks. Phase three will involve the placement of candidates in foreign groups to complete tasks that demonstrate how well they will function in challenging circumstances with others. It will include international classes of three candidates in round four. It will demonstrate Their ability to work together in conditions like those in the moon colony (Musson, Sandal, Gro. M, & Helmreich, 2018). The crew will be mainly composed of specialists such as agronomists, medical researchers, and software engineers. However, all group members will be trained, fields such as medicine to ensure that everyone can take responsibility in an emergency. They will have to be transported to the colony after the crew has been identified and trained.

The Lunar Exploratory Mission will represent design progress and stimulate the economy, creativity and international cooperation and ensure that humanity evolves further. Early in the next century, it will also finance a human-crewed Mars expedition and lower its cost. Thus, small-scale self-sufficiency will be feasible by maximizing a community's total output ability, even in such an extreme setting as a colony on the lunar surface. Building loads on the Moon are very different from Earth, with the dominating internal strain, and the most significant factors in construction are radiation safety and airtightness. Consequently, like caves or mines, underground solutions would be the most appealing. For these to also be used in recreation, it is necessary to design the colony interior to give an impression of the outside agricultural sections. Current design theories and analytical techniques form a sound foundation for developing a lunar base from which it can make modifications.

References

Astronaut selection and medical requirements. (2009). *Lunar Outpost,* 151–182.https://doi.org/10.1007/978-0-387-09747-3_6

Astronaut Requirements. (2015). Retrieved February 8, 2021, from NASA website: https://www.nasa.gov/audience/forstudents/postsecondary/features/F_Astronaut_Requireme nts.html

An Overview of Lunar Base Structures: Past and Future | AIAA SPACE Forum. (2012). RetrievedFebruary 8, 2021, from Aiaa.org website: https://arc.aiaa.org/doi/10.2514/6.2002-6113

Benaroya, H. (2006). Lunar base design: a paradgm for the capstone design course. *IEEEAerospace and Electronic Systems Magazine, 21*(7), S_33–S_36. https://doi.org/10.1109/maes.2006.1684276

Benaroya, H., Bernold, L., & Chua, Koon Meng. (2021). Engineering, Design and Construction ofLunar Bases. *Asce.org*, 229–242. Retrieved from https://cedb.asce.org/CEDBsearch/record.jsp?dockey=0136788

Colwell, J. E., Batiste, S., Horányi, M., Robertson, S., & Sture, S. (2007). Lunar surface: Dustdynamics and regolith mechanics. *Reviews of Geophysics*, *45*(2). https://doi.org/10.1029/2005rg000184

Duke, M. B., Mendell, W. W., & Roberts, B. B. (2015). Strategies for a Permanent LunarBase. *Lunar Base Agriculture: Soils for Plant Growth*, 23–35. https://doi.org/10.2134/1989.lunarbaseagriculture.c2

Maiwald, V., Vrakking, V., Zabel, P., Schubert, D., Waclavicek, R., Dorn, M., ... Zeidler, C. (2020). From ice to space: a greenhouse design for Moon or Mars based on a prototype deployed in Antarctica. *CEAS Space Journal*. https://doi.org/10.1007/s12567-020-00318-4

Moon's South Pole in NASA's Landing Sites. (2019). Retrieved February 8, 2021, from NASAwebsite: https://www.nasa.gov/feature/moon-s-south-pole-in-nasa-s-landing-sites

Musson, D. M., Sandal, Gro. M, & Helmreich, R. L. (2018). Personality Characteristics and TraitClusters in Final Stage Ast...: Ingenta Connect. Retrieved February 8, 2021, from Ingentaconnect.com website: https://www.ingentaconnect.com/content/asma/asem/2004/00000075/00000004/art00007

Ruess, F., Schaenzlin, J., & Benaroya, H. (2006). Structural Design of a Lunar Habitat. *Journal ofAerospace Engineering*, *19*(3), 133–157. https://doi.org/10.1061/(asce)0893-1321(2006)19:3(133)

Office of the Surgeon General (US, Center for Mental Health Services (US, & National Institute ofMental Health (US. (2001, August). Chapter 2 Culture Counts: The Influence of Culture andSociety on Mental Health. Retrieved February 8, 2021, from Nih.gov website: https://www.ncbi.nlm.nih.gov/books/NBK44249/

Smith, A. (1993). Mechanics of Materials in Lunar Base Design. *Applied MechanicsReviews*, *46*(6), 268–271. https://doi.org/10.1115/1.3120354

Yibing, Z., Ronghui, Z., Linhui, L., Lie, G., & Mingheng, Z. (2014). Walking Wheel Design forLunar Rove-Rand and Its Application Simulation Based on Virtual Lunar Environment. *Advances in Mechanical Engineering*, *6*, 389272.https://doi.org/10.1155/2014/389272

Chapter 5. Semi-Autonomous Mars Surface Robotic Exploration Mission Proposal (SMSREM)

Abstract

This report proposes the Semi-Autonomous Mars Surface Robotic Exploration Mission (SMSREM) to enhance human potential, augment physical safety by providing means for performing several tasks without requiring an immediate human presence, and assigning tasks from the ground to crews typical for mission control. As NASA's Vision for Space Exploration calls for human beings to return to the Moon and Mars for eventual human exploration, the difficulty of such various tasks would involve a never-ending use of automation and robotics to assist human crews (Mishkin, Lee, Korth, & LeBlanc, 2007). This paper considers a cooperative robotic device consisting of a copter and a rover for autonomy during space exploration missions. The Copter's purpose is to explore a new environment to optimize information about a scientific task, which the Rover will carry out in linear time logic (Nilsson et al., 2018). A suspension device covers the instruments inside the robot on wheels. It can prolong the life of the robot and avoids physical harm to the tools inside. This robot has an external configuration for exploring craters. The experimental results support the proposed solutions and show platforms' capacity to explore environments with unusual soil shapes and obstacles of various dimensions.

Introduction

The SMSREM Mission 2025 will expand on past discoveries, and its integration of Rover and copter, rovecopter, is the crucial next step in the strategic program for Mars exploration. Color, multi-spectral, and stereo imaging systems for observation, necessary for addressing the Mission's scientific objectives, involve statements with the ability to perform side and stratigraphic surveys and multi-space scale studies on several goals and to support the navigation of rovers. Imager device project functions involve the geological background on the rover route to choose further research and sampling employing arm-mounted instruments. The performance of these arm-related activities is documented and validated (Mustard et al., 2013). The primary data sets needed to map and interpret the geology, the chemical industry, and mineralogy of the planet in terms of geology and environmental history of Martia were gathered in several very successful missions at the global, regional, and local level (e.g., Malin and Edgett 2001; Squyres et al. 2004; Grotzinger et al. 2014; Ehlmann and Edwards 2014). Their predecessors' findings influenced several of these new missions, experiments, and investigations, and some of these missions continue to operate to date. The SMSREM aims to perform independently multi-solar navigation without human intervention to increase efficacy and return to science. The primary aim of SMSREM is to enhance both autonomy and productivity by developing principled methods that define the most relevant areas tobe explored in a specification-guided manner.

Semi-Automated Robotic Mars endeavors will introduce multiple challenges to the current approach for the mission operations. It is an endeavor that will affirm the potential for humans to leave our home planet and advance outward into the cosmos. While it will be a tiny step on a galactic scale, it will be a big one for humans, as it will take a minimal return capacity to leave Earth (Hoffman et al., 2021). A surface investigation by SMSREM will incorporate a broad scope of activities. Counting, Observing, and studying the geology, chemistry, and function of the surfaceand subsurface of the Martian atmosphere, collecting and analyzing samples in the outpost laboratory, and conducting tests intended to test the human ability to inhabit Mars (Hoffman et al.,2021). The following question suggests fulfilling the objective to find life on Mars. Is there life under the oxidized regolith layer on Mars soil? Can we change the atmosphere of Mars? Moreover, How is Martian Environment? Is it suitable for human life? If proof existed that life had evolved elsewhere in the universe and might expand, it would have significant scientific consequences. Some claim that such a discovery is one of the best scientific discoveries ever. Scientists have discovered life deep at the bottom of the ocean around thermal winds and bacteria in ice shelves at the Earth's poles in the past decade. We continue to discover life where life did not exist before it was thought. The existence of water is invariably a common thread. Some claim this supports the conclusion that liquid water is abundant everywhere life evolves. Others argue that the transformation in a hostile environment of prehistoric Earth of purines and other beginning compounds into real-life forms was a fortunate occurrence that cannot be replicated (Brien, 2003). The discovery of any signs of alien existence will eventually resolve this matter. The possible discovery of millions of dollars in the pursuit of extraterrestrial intelligence was so amazing. Even with its uncertain conclusions, it was not only the probability of meteorite-related Mars existence that contributed substantial government financing to the Mars sample return mission. The return of a sample that contains evidence that the life of Mars began to grow will not doubtless lead to even more desire to speed up the Mars plan, or even the waterfront case, which eventually offers the impetus to plan a human-crewed Mars mission.

Science Goals of Autonomous Mars Surface Robotic Exploration Mission:

Primary Goal

To detect, classify and classify indications of life under oxidized regolith layer on Mars soil, present or extinct.

To identify and analyze Martian rocks and soils that are most likely to preserve signs of ancient geological environments and future conditions.

Secondary Goal

To implement robotic exploration technologies to change the Martian climate in the nextdecade or more.

To provide meteorology and Martian climate environmental data to support future Marshuman and robotic missions.

To keep the promotion of new students in scientific and engineering education in scienceand technological advancement.

To offer mobile robotics research application emphasis with significant applications intechnology transfer.

Mission Users and Operation

The value of this Mission lies in the ongoing quest for life on Mars. Researchers can discover solutions to their problems by searching beneath the surface using a new life quest. One huge concern is whether Mars has always been habitable. This challenge is a task for all since both the past and the present are discussed. To discover the surface of Mars, the rovers work in conjunction with a team of Earth scientists and flight control engineers to prepare and analyze the success of regular mission operations for each Rover (Tunstall et al., n.d.). As a whole, this SMSREM mission represents distributed automated robot system for semi-autonomous planetary surface exploration. The SMSREM system is a complex system divided into two teams of Earth's human aircraft, regularly engaging on Mars with the two rover systems. A team that performs the uplink command sequence and a group that conducts downlink telemetry analysis are part of each flight control team. A couple of scientists direct the uplink and downlink teams to accomplish the project's aims and goals through the discovery process.

Furthermore, the Uplink Team is split into smaller teams that organize rover operations and develop supporting commands that control rovers' everyday activities on the Martian surface. The Downlink team is divided into smaller sections that track the regular sequence of controls, rover performance, and secure and efficient rover activity. These smaller teams are grouped into concentrating disciplines of spacecraft and rover design by a subsystem of engineering.

Science is closely linked to the Mars Exploration Program to determine the extent to which Mars provided necessary conditions for the prebiotic compound formation and conservation and whether life began and evolved. This goal can be broadly described to define Mars' habitability and understand the role of tectonic and climatic processes in providing life-changing conditions. It is of fundamental importance that water is found and its interaction with crustaceans. Therefore, three of the MER objectives are focused on exploring past water evidence: (1) researching landing sites that are highly likely to contain evidence of liquid water behavior, (2) searching for and characterizing a range of rocks and soils that contain evidence of past water activity, and (3) extracting evidence of environmental conditions when liquid water has been preserved(Golombek et al. 2003). The other scientific goals of SMSREM concern the Mars Exploration Program aimed at defining the type and sequence of the geological processes which produced and changed the Martian crop and surface: the spatial distribution and composition of the minerals, rocks, and soils surrounding landing sites and determining the nature of the local geological surface phase, Calibrating and testing orbital remote sensing data and establishing the volume and size of iron-containing minerals in each site for landing, classifying, calculating the relative amounts of various $H2O$ or OH mineral forms, or being aqueous process forming markers, characterizing mineral assemblies and textures of different rock and soil types and characterizing m m m These are primary field geological goals that can be achieved at any landing site, but which are the groundwork for the first three objectives relating to past waters and hence habitability.

Stakeholders

Stakeholders involved in this Mission ranges from the top branches of the US governmentto the project execution level both inside and outside of NASA, resulting in diverse perspectives, interests, and influences. Different Stakeholders can raise new issues as they might have radically different perspectives, motivations, and drivers. Therefore, The determination of the wanted result relies not on the perception that it is consistently correct but on the stakeholder's point of view. Because the initial challenges and associated resource requirements are not clear, they are often overly restricted, cannot be definitely overcome by critical thinking, and may require creative solutions. SMSREM has set priorities to ensure that priority production of value flows is provided to its stakeholders so that the full benefit is returned to SMSREM, thereby ensuring a viable discovery enterprise for more scientific discoveries. To integrate into campaigns that deliver value, the physical and information systems that comprise the underlying infrastructure for exploration must be sufficient. From a management viewpoint, this implies that it would be beneficial to establish a precise position (and even organization) to ensure that stakeholder value is safeguarded and effectively maintained during the initiative's infrastructure process (Balint & Stevens, 2016).

Mission Statement

The SMSREM is a technology demonstration intended to show the feasibility and utility of helicopters for Mars exploration. Numerous people will benefit from this project.

The scientific community is one of the beneficiaries of the Mission. Biologists will be well acquainted with the past, present, and future life-sustaining capabilities of Mars and the potential organization. Geologists can determine whether fossils are just a phenomenon occurring on Earth or whether similar processes occur elsewhere in the solar system. The scientists working on future human- crewed missions to Mars may know more about the potential for life to grow on Mars and how it could affect humans and the Martian environment. Humans the world over will also benefit from this knowledge, especially since a question that intrigues people globally is whether life exists on Mars. The benefactors also add importance to the Mission. This Mission will collect data on the mission subject, Mars, to accomplish each part of the task. The Mission will use technologies from a potential NASA mission, the Rovocopter, to discover life under the earth. On this mission, SMSREM will make use of the X-Ray Spectrum Alpha Particle (APXS), to analyze the abundance of rock components and to scan for life blocks in the house. Another form of technology that willbe sent to Mars to look for life is an 8-foot biological test shaped like a torpedo; looking for life on Earth's surface is the subsurface explorer for assessing life (SEAL) (Villard, 2013). These are indispensable in Mars' quest for existence. The Rover is fitted with a microscope to extend the fossils beneath the Martian soil and take photos of these samples. Another form of technology is being sent to Mars, now used on Earth, to seek life. Enceladus Explorer (EnEx) is an interplanetary mission with a maneuverable ice melting sonde on its surfaces. These are indispensable in Mars' quest for existence. The Rover is fitted with a prism to amplify and take pictures from the samples, to detect fossils under the Red planet. APXS will send the photos back to the scientists who will investigate the existence of incredibly tiny fossils.

Classes or volunteers can also review visits to photos to help researchers detect potential fossils from the vast quantities of images sent. MOMA will again be added to the future part of this Mission. The site designed to resemble the Earth would scan for the organic molecules. An ultraviolet miniature microscope will also be tested for detection of bioluminescence (Villard 2013).

Accurate Positioning in Natural terrain is also going to be used in this Mission. The goal is to allow the Rover to maneuver in the rough terrain reliably for more extended missions. The problem of rover positioning is broken down into three additional areas of study, based on the necessary accuracy and natural framework for the definition of rover commands. These three areas are absolutely landscaped over a wide range of long cross-sections between scientific sites; excellent absolute positioning in the narrow field of the research site; and accurate relative positioning between the Rover and the scientific target of terminal maneuvering and instrument placement. The Copter is the significant component of the SMSREM rover during the flight to Mars and partially plied the landing gear. The helicopter is deployed in an area considered appropriate for helicopter operations after the Landing on Mars, and a safe distance is taken away by the Rover. The aircraft shall perform a sequence of independent flights approximately 1 minute after deployment. Before descending and landing again, the rovocopter shall take-off during each aircraft, climb, float, and optionally cross between different route points. Although the rovocopter is fully autonomous during flight, the waypoints are determined from Earth before the flight. Weather conditions are monitored from Earth on Mars, and flights are ordered only when sufficiently benign conditions are expected to occur. As this technology demonstration did not prioritize the prevention of landing hazards, each flight will be commenced and terminated in a pre-inspected area determined to be safe concerning obstacles and terrain pitch.

Mission Constraints

There are significant challenges that must be overcome to complete the SMSREM mission. Owing to the delays in communication with remote spacecraft, deep-robot missions have exceeded saved sequences' execution. A single command load on certain flights may consist of tens of thousands of orders, weeks, or months tracking spacecraft's behavior. The command sequences must be validated closely to ensure that their execution is consistent with the human operators' original intent and that there should be no risk to a spacecraft since these spacecraft must be run without direct intervention.

Any control sequence leaving the vessel idle or unintendingly disconnecting the critical system will cause the spacecraft's Mission and malfunction unnecessarily. Traditionally, the issue has been overcome by two methods: rigorous sequence evaluation, mostly involving high-faith sequence checking, and built-in spacecraft device on-site fault recovery. If no instructions have been placed on the spacecraft during a specific time, the spacecraft can autonomously change its actions to point its antenna to Earth so that new orders are best obtained. If the spacecraft is powered by solar radiation, it will target the sun's arrays to ensure hope and will thus not perish in the awaited future. The consequences of environmental problems such as cosmic ray impacts on electronic elements are also mitigated by online fault reactions, which in some instances will respond where some other event prevents the output of the sequent. A project failure's tremendous effect has contributed to adopting a strongly conservative method for designing and validating command series. Time tags are used with typical commands to signify precisely when to activate them. The creation of a command load may take weeks or longer to establish a conflict-free operation plan and construct actual sequences of commands. The atmosphere is likely to influence the spacecraft in unpredictable ways for orbiters and free flyers, so constructing commands are not troublesome long before use.

Although robotic deep-space missions have so far succeeded in utilizing techniques that have made exploring the solar system feasible, robotics systems still have to be modified to become efficient and productive on an astronaut-size time scale when they reach Mars. This work advances and validates key technology to allow planetary rovers to cross long distances safely and autonomously on robust terrains. New strategies for measuring the landscape in real-time can be developed by relying on the measurements made by on-board sensors for the physical properties of the soil (such as inclination, roughness, and hardness). In the field of navigation, the logic rovers are used continuously to guide the Rover to the safest and more traversable terrain, thereby going towards the intended target and avoiding unforeseen obstacles. New techniques for rover navigation are being developed. Both in-field evaluations and in rover navigation, rover survival is optimized in many resolutions. Several field studies will be performed on representative natural terrain for assessment, and the algorithms for sensing and navigation will be improved based on the experimental findings achieved. The results of this research will improve rover safety dramatically and enable robotic exploratory missions to complete in high-risk fields.

Landing Site

To ensure that the proposed data is obtained and achieve optimal success, landing sites must be carefully chosen. For the SMSREM Mission, Meridiani Planum will consist of a lowland region of the heavily cratered mountains at the East side of the western hemisphere of Mars. Meridiani Planum is situated on the westernmost region of Terra Meridiani in the direct 2 degrees south of the Mars equator (centered at.2°N 357.5°E). It involves a particular case of gray crystalline hepatitis. The area mapping shows the north-westerly valley networks, down the topographical gradient generated by Tharsis's bending [Phillips et al. 2001]. There were substantial deforestation and denudation in the region [Hynek and Phillips 2001; Parker 2002], continuing until late Noachian. The image was taken by 128 Gridding pixels per degree of data from the Mars Orbiter Laser Altimeter. Detected craters vary from comparatively young craters such as Vega, Viking, and Framto heavily soiled and infilled craters like Eagle and Vostok.

HiRISE image of the Martian crater Eagle – the dot in the

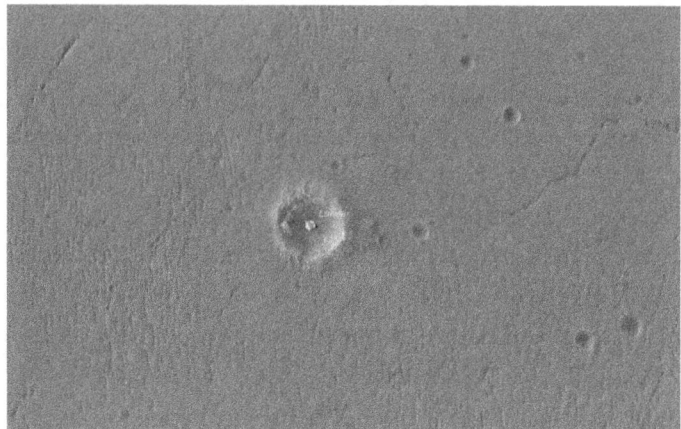

middle of the crater is the lander.

They document the gradual erosion by eolian of the delicate sulfate basement and basaltic sand infilling [Grant et al., 2006a]. The number of these craters, including those <250 m in diameter thatin orbital photos are distinctly scant (Figure 2), indicates a late Amazonian [Lane et al., 2003] gross surface age of basaltic sand and granular wax. Furthermore, the contrast with Hesperian era surface areas like Viking Lander 1 and 2, Mars Pathfinder, and Gusev of the estimated crater density at Meridiani planum reveals considerably fewer craters. The deficiency of Meridiani craters claims that the entire Hesperian cratering record has been eradicated, further evidence of an erosion of the former Noachian craters on Meridiani planumand layered terrains as a whole (Malin and Edgett, 2000a; Edgett, 2005). In addition, Meridiani Planum was specifically picked for SMSREM as Remote sensing on NASA's Mars Global Surveyor from orbit by the Thermal Emission Spectrometer (TES) revealed that sections of Meridiani's planum contain as much as 20% crystalline surface hematite. Hematite is an iron-oxide-mineral, and is mostly associated with liquidwater and is a grey crystalline variety on Earth (PASCHKE & HEARST, 2015).

Concepts of Operation

The Mission will be launched by Ariane 6 from NASA Langley Research Centre. The Mission provides the surface mission communications relays covering the SRL entry, descending and deplaning, surface operations before the starting time of the MAV. It will be launched at the Langley Research Center in April 2021. Ariane 6 is expected to land on Mars on 17 November 2021. On Mars, in the explicit 2 degrees South of Mars' Equator (centralized at.2°N 357.5°E), it will land on the Meridian Planum site. The landing region is to be explored by it for at least 2 March years (4 Earth years). Data collection will take at least four years because it takes time for the surface to dig deep, and numerous samples are taken. One Rover that can also fly as a rovocopter is sufficient to dig and act as a subsurface explorer.

Scope Summary

Need: To understand the past, present, and future of Mars to seek life, consider the surface and the history of the Earth, and prepare for future human exploration outward into the cosmos.

Goal: Explore a geologically diverse landing site to detect, classify indications of life underoxidized regolith layer on Mar Soil and analyze Martian rocks and soils that are most likely to preserve signs of ancient life.

Objective: To search for fossils of living organisms such as primitive single-celled organisms.

Mission: Safely Transport rovocopter to Mars in the quest for ancient life.

References

Arena, P., Di Giamberardino, P., Fortuna, L., La Gala, F., Monaco, S., Muscato, G., … Ronchini,R. (2004). Toward a mobile autonomous robotic system for Mars exploration. Planetary and SpaceScience, 52(1-3), 23–30. https://doi.org/10.1016/j.pss.2003.07.002

Drake, B. G. (2009). Human Exploration of Mars Design Reference Architecture 5.0 Mars Architecture Steering Group NASA Headquarters ACKNOWLEDGEMENTS. Retrieved from https://www.nasa.gov/pdf/373665main_NASA-SP-2009-566.pdf

Ellery, A., Ball, A. J., Cockell, C., Dickensheets, D., Edwards, H., Kolb, C., … Richter, L. (2005).Vanguard—a European robotic astrobiology-focussed Mars sub-surface mission proposal. Acta Astronautica, 56(3), 397–407. https://doi.org/10.1016/j.actaastro.2004.05.069\\

Flight Control System for NASA's Mars Helicopter | AIAA SciTech Forum. (2019). RetrievedFebruary 21, 2021, from Aiaa.org website: https://arc.aiaa.org/doi/abs/10.2514/6.2019-1289

Kalita, H., Gholap, A. S., & Thangavelautham, J. (2020). Dynamics and Control of a HoppingRobot for Extreme Environment Exploration on the Moon and Mars. 2020 IEEE Aerospace Conference. https://doi.org/10.1109/aero47225.2020.9172617

Mishkin, A., Lee, Y., Korth, D., & LeBlanc, T. (2007). Human-Robotic Missions to the Moon andMars: Operations Design Implications. 2007 IEEE Aerospace Conference. https://doi.org/10.1109/aero.2007.352960

Toward Specification-Guided Active Mars Exploration for Cooperative Robot Teams -MurrayWiki. (2018). Retrieved February 22, 2021, from Caltech.edu website:

http://www.cds.caltech.edu/~murray/wiki/index.php?title=Toward_Specification-_Guided_Active_Mars_Exploration_for_Cooperative_Robot_Teams

Trebi-Ollennu, A., Baumgartner, E. T., Leger, P. C., & Bonitz, R. G. (2021). Robotic Arm In-Situ Operations for the Mars Exploration Rovers Surface Mission. 2005 IEEE International Conferenceon Systems, Man and Cybernetics. https://doi.org/10.1109/icsmc.2005.1571409

Tunstel, E., Maimone, M., Trebi-Ollennu, A., Yen, J., Petras, R., & Willson, R. (n.d.). MarsExploration Rover Mobility and Robotic Arm Operational Performance. Retrieved from https://www-robotics.jpl.nasa.gov/publications/Mark_Maimone/MobIDDPerf90sols.pdf

Starek, J. A., Açıkmeşe, B., Nesnas, I. A., & Pavone, M. (2015). Spacecraft Autonomy Challengesfor Next-Generation Space Missions. Advances in Control System Technology for Aerospace Applications, 1–48. https://doi.org/10.1007/978-3-662-47694-9_1

Using a Robotic Helicopter to Fuel Interest in and Augment the Human Exploration of the PlanetMars | AIAA SPACE Forum. (2012). Retrieved February 21, 2021, from Aiaa.org website: https://arc.aiaa.org/doi/abs/10.2514/6.2003-6275

Chapter 6. Challenges and Risks to Human Spaceflight – Risk Management Plan Technical Report

Abstract

This paper examines the challenges and risks and projects the implications and the forward work necessary to ensure the success of Human Mars missions. Human spaceflight travel poses new challenges to conventional approaches to risk management. From the human onset, spaceflight hasbeen regarded as an inherently hazardous operation (Langston, 2016). In terms of distance to traveland length of the mission, human spaceflight will present specific psychological obstacles that arenot comparable to any other human endeavors that have ever been attempted (Manzey, 2004). Thispaper provides integrative ethical research, and an ordinance to support risks faced by radiation, nutrition, and breathable air to facilitate the making of prudent decisions and risk assessments forspaceflight activities. The selection of risk presented to evaluate the system's supportability approach is discussed, and their potential effects on the supportability metrics and the various configurations between them.

INTRODUCTION

Unlike the early spaceflights, which only lasted for a few hours or a day, the current spaceflights are marked by much longer length and more crew-heterogeneous cultural and professional background (Marov, 2020). Astronauts will travel farther from Earth and remain in orbit for longer periods, without the resupply from Earth or the ability to abort with the emergency. This emphasizes the need to evaluate and explain social perspectives on topics of risk, ambiguity, and standardization (Manzey, 2004). Astronauts participating in a long-term spaceflight missions suchas Mars require particular radiological safety technology because of the difficulty of the radiation system and of their resistance to hazardous and energy particles like neutron and other radiation that are highly ionizing (Sihver, 2019). The risk of cancer in astronauts has been significantly increased with long-term periods of increased spatial radiation doses. Further potential exposures to cosmic radiation entail skin flexible schedule, CNS injuries and acute ray disease syndrome (Mortazavi, 2019). This paper addresses health risks of space flights for astronauts, the limitations and constraints of passive thermal insulation, the need for radiosensitive monitoring and the use of medical measures. Vitamin C is particularly explored in the paper as a compelling not harmful, economically efficient and easily obtainable radiation mitigator (Chambial, Dwivedi, Shukla, John,& Sharma, 2013).

Long Human Spaceflight to a planet such as Mars requires greater propulsion capacity to deliver missions as the distance from Earth rises and planetary physics renders daily supplies unfeasible (Foust, 2019). The provision of food and nutrition is thus another major challenge involved in human spaceflight missions. Designing a food system for Mars missions would be much more complex than establishing one for the International Space Station (Douglas, Zwart, &Smith, 2020). While numerous food classifications exist on Earth, their capacity to achieve spaceflight demands has yet to be determined, and their feasibility within cost and schedule constraints remains unknown, particularly since the Mars mission requires crews to travel at breakneck speed away from their home planet for six months (Douglas, Zwart, & Smith, 2020). This paper highlights the tremendous challenges of developing a space exploration food system that is often ignored in the run-up to Mars. Also if rocket and technology systems work flawlessly, the journey will end in the same tragic manner when the food system meets no such requirement as other missions never even left home but were wrong because of deficiencies in food and feedingsystems.

As the length of space flights expands, it is highly important to retrieve useful consumables from waste materials (Bourland & Smith, 1991). Because of its sensitive nature, the limited volume, and almost total closing of the enclosed space, the air quality subsystem is the most crucial among the subsystems that are present within a life support device. There are various causes of air pollution in a closed cabin spacecraft system. The major pollution source within a spacious cabin is the results of human or animal absorption, irrespective of its duration or design(Mulloth & Finn, 2021). Chemical leakages into the cabin environment caused by the continued use of instruments or equipment for research and live-support systems add additional loads to the air revitalization system. (Boyada, 1986), Failure of gravity and atmosphere are essential differences between space and the Earth. The selection of air pollution control equipment and processes for the spacious cabin has been made more difficult because of serious problems (Mulloth & Finn, 2021). The benefit is that land systems are identical to fundamental consumables, such as air and water. There is a chance of air leakage due to the small pressure or vacuum on the spacecraft. This paper presents a broad picture of aviation systems, spacecraft problems and addresses the most critical and resource-consuming function of carbon dioxide removal technology and other rehabilitation activities, containing carbon dioxide reduction, oxygen production, trace pollutant management, and other subsystems.

CAUSES AND RISKS INVOLVED IN RADIATION

In the harsh climate, a deep space mission presents many risks to space crews. In addition to technical and operational problems, because of microgravity that affects brain function and cardiovascular system, weight bearing systems, immunologic responses and so on, the crew will also be physiologically changed (Demontis et al., 2017). In addition, due to isolation, confinement and alienation from the earth, there is a lack of privateness, social tensions and psychosocial tension (Loades et al., 2020). Furthermore, the crew may have very different backgrounds, nationalities, socioeconomic, cultural and ethnic values. When dealing with many health issues, one must also take into account the fact that medicines on the body under space conditions can function differently (Sihver & Mortazavi, 2019). The most important challenge is to exposure of space crews to high energy ionizing radiation, despite all the problems mentioned above. From 1950, the casualties of the Hiroshima and Nagasaki Nuclear Bombings were 130,000 people. A Life Scale Research (LSS) has been initiated to track these survivors and track their wellbeing for the short or long term. (Cucinotta, Durante, 2009). As the explosions of the bombings spanned the beginning of the research during five years, for that timeframe, there is an evidence gap in the cancer rate. This duration is not thought, except for leukemia, to have a significant effect on awareness of most cancers (Jones, 2010). Despite this information gap, the doses obtained by bomb survivors (from between 50 and 2000 mSv) are significantly superimposed on the anticipated doses of long-distance NASA missions. During the LSS, the risk measured for the occurrence of cancer is very nearly linear with the dose got (Jones, 2010).

High energy charged particles, and in particular galactic cosmic rays (GCRs) containing high- energy (high-E) nucleus with heavier (high atomic Z) components, make up the portion of space radiation of concern ("HZE particles") (Onorato, Di Schiavi, & Di Cunto, 2020). This sensitivity to radiation leads to two main types of radiation hazards: Short-term consequences of relatively high levels of radiation, such as a solar particle event (SPE) (Chancellor et al., 2018). This form of radiation risk is mainly due to fragile tissue depletion, such as bone marrow, intestinal epithelia, skin, etc., which can lead to symptoms that affect the well-being and performance of the teams at work. Long-term exposure to expected solar and galactic cosmic rays results in increased cancer probabilities and, even, alteration in brain cells, reproductive bodies, or other tissues. (Barbrow, 2020). There are no proven ways to eliminate the risk of space-light radiation in people living and working in spacecraft, planetary surfaces, or space stations. The aim of shielding radiation in space is therefore to reduce these risks to a relatively safe level. (Szocik, Norman, & Reiss, 2019). In order to accomplish this objective, threats need to be identified, in particular as to whether or not space radiation poses a danger of not being exposed to Earth's common sources of radiation, i.e. primarily x-rays and similar LET radiation. Secondly, it is important to predict the magnitude of these risks so that procedures can be developed on the basis of quantitative information (Furukawa et al., 2020). Such estimates must also be accurate in such a way that the countervailing designs do not incur unnecessary costs or that the underestimation poses unreasonable risks.

CAUSES AND RISK INVOLVED IN NUTRITION

Food security in the orbit remains a major issue. Therefore, food manufacturing and packing are strictly regulated by the preflight to ensure that the pollution level meets the requirements enforced for spatial foods (Mord, 2009; NASA, 2006; Perchonok & Douglas, 2008). Using techniques like thermostabilization and radiation ensures a safe food supply and decreases the chance of microbial contamination (Douglas et al., 2016). However, viable microorganisms can also be found in the food. While commercially sterile products including changed form and irradiated foods have not been microbiologically checked, consistency and swelling are tested for packaging. Both commercial products are purchased from the same lot, and every space in the package has a bar-coded serial number that can be traced from the original lot (Lupo, 2015). If any commercial food items are recalled, NASA will automatically gain awareness of the contaminated lots and will alert astronauts to not consume the food. Since spacecraft is a confined area in which air, water, and food sources are processed (Schwartzkopf, 1992), there is still the possibility of microbial bridge of substances just after launch. Crew and passengers are also microorganism reservoirs. Most amove of environmental or human origin are not serious health risks, but can pose a health risk to astronauts whose gut bacteria are compromised (Chouker, 2011). Thus, the overall load of microbial cells must be monitored, because spaceflight is an enclosed universe in which air, water, and nutrient supplies are filtered (Schwartzkopf, 1992), and there is also the risk of bacterial cross-contamination of food after the mission. Crew members are still repositories of microorganisms. Most adverse environmental and social microorganisms are not hazardous to healthy people; they pose a danger to astronauts withweakened immune systems (Chouker, 2011).

In order to ensure effective standard of health for crews, a cumulative load of local species must also be tracked. Microbiological water contamination is also a major health issue (Berry, Xi, & Raskin, 2006). Biofilms are one of the main factors for growing the persistence of antiseptic microbiota in the storage and treatment of drinking water (Van Houdt & Michiels, 2010). If water is supplied to rehydrate indoor or indoor food and beverages, polluted water also may cause foodborne illness on space flights (Kim & Rhee, 2020). Reasonable monitoring anddecontamination is also essential to reduce the risks to crew health. Space travel research was initiated during Mir's program to develop techniques for collection for a long-term space mission and was performed periodically on the ISS to control contamination (Hurst, 2017). The ISS has identified acceptable limits for pathogenic bacteria in the air, surfaces and drinking water. Despiteefforts to monitor the microbiological status of the spacecraft, however, viable microorganisms have in the living activity and crews been identified from air, surfaces and liquids (Checinska Sielaff et al., 2019).

CAUSES AND RISKS INVOLVED IN BREATHABLE AIR

From the beginning of crewed space travel, maintaining appropriate safe cabin air has been a challenge. Beyond preserving cabin temperature and humidity levels within safe and comfortable limits, the key issues are carbon dioxide partial pressure, trace chemical contaminant concentrations, and particulate matter accumulation (Jernigan, Gatens, Joshi, & Perry, 2018). As the range of crewed space exploration missions lengthens and the mission's targets expand beyond low Earth orbit, advances in air purification device mass, control, volume, and logistics are needed. Air purification technologies focused on adsorption and catalytic oxidation have proven to be effective; however, process inefficiencies and continued dependence on expendable capital pose obstacles to expanding the frontiers of space exploration (Perry et al. 2002). Efforts have been identified and growth efforts are progressing to address these ISS and beyond inefficiencies. Air quality management technologies for the next generation of aircraft are being developed for the advances in regenerative adsorbent media and oxidation processes (Manisalidis, Stavropoulou, Stavropoulos, & Bezirtzoglou, 2020). Besides the wide spectrum management of air quality in a limited volume, low power kit, this future generation of the air purification processes needs to provide a factor of up to three reductions in logistics mass (Perry et al. 2002).

POSSIBLE SOLUTIONS FOR RADIATION

In the form of interaction with local environmental partitions with the shield's constituent atoms and nuclei, the potency of each Shield substance is determined by the conveyance of energetic particles within the shield (Martinez & Kingston, 2012). For different materials, these interactions vary significantly. High-hydrogen-content materials are usually more effective in protecting spatial radiation shields, but often lack qualities that give the vehicle or habitat the requisite structural integrity (Cohen 2004). The most important aspect of space exploration is protection. Safety is paramount. It is evolving in response to the growing difficulty of space traveland the steady progression of human missions to conduct manned missions to the Moon, Mars, and beyond. On this perilous course, we will encounter a slew of issues, including a slew of unknowns and natural-source threats in outer space, all of which will coexist with great technologies and breakthroughs (Martinez & Kingston, 2012). Among the natural issues, the paper analyzes cosmic radiation as the most dangerous factor of space exploration.

While some recent papers have focused on traditional radiation shields, research indicates that radiation shielding may be more important during space missions (Chancellor, Scott, & Sutton,2014). Through the significant radioprotective effect of vitamin C, this antioxidant agent is introduced as an efficient, non-toxic, cost-effective, and easily accessible radiation protector for lifestyle-threatening radiation (SPE) exposure (Chancellor, Scott, & Sutton, 2014). Vitamin C is a promising radiomitigator that can be used, before 24 hours after exposure to high levels of ionizing radiation in the event of solar particulate events.

The long interval will thus allow astronauts to not only evaluate their radiation exposure before deciding on any therapeutic action but also to consult Earth's specialist doctors to double- check vitamin C and other therapeutic agent dosages (Chambial, Dwivedi, Shukla, John, & Sharma, 2013). Furthermore, in light of concerns about the effects of space radiation on astronauts' eyes, it was recently discovered that the use of vitamin C is inversely related to the risk of cataracts. Researchers have discovered a connection between ascorbate serum, nuclear, and subcapsular cataracts in the opposite direction. Improved insulation for earth radiation workers usually provides additional radiation protection (Pippa Wysong, 2006). Unfortunately, space shielding is problematic, especially when GCR is taken into account. High-energy radiation is very penetrating: thin to intermediate protection decreases exposure, but as the thickness of the protection increases, the shielding performance deteriorates (Wilson et al., 1997). This is the result of the formation of a large number of secondary particles, including neutrons, as a result of GCR's nuclear interactions with the shield. These particles have less energy than primary interstellar particles but may have higher quality influences (Norbury et al., 2016). Furthermore, shields obviously pose a problem for a large number of people. A large load that is added solely to minimize radiation exposures is subject to a substantial mass starter penalty, which could dramatically increase the mission cost (Chambial, Dwivedi, Shukla, John, & Sharma, 2013).

Typical wall shielding is around 5 g/cm2 Al, but due to numerous payloads and racks on the ISS, the effective shield is near 20 g/cm2 Al in a variety of areas (Zeitlin et al., 2019). All protons with energies of less than 100–200 MeV will be avoided by this Al dense. As a result, it works well for trapped and most SPE-specific radiation (Durante and Cucinotta, 2011). Protons with energies greater than a few MeV can pass through the skin and be stored in the inner organs. The spacecraft which have a storm cover, which is a small region with a thick shield (>20 g/cm2 Al), to protect it from large satellites. The SPE persists for less than two days in the majority of cases and just a few hours in the most serious cases. During this limited time, the crew will move in this small area to defend themselves (Motoba et al., 2020).

POSSIBLE SOLUTION FOR NUTRITION

With proper nutrition, crew wellbeing can be preserved during and after long-term spaceflight. To improve food intake by crew members, a number of foods with positive sensory features, as well as sufficient meal preparation and consuming time are used (Enrico, 2016). In addition to the microgravity, constant light exposures are a frequent cause of anorexia, variations in dietary intakes, neurotransmitter monoamines, and peripheral hormones suggest. Deprivation of sleep may have an effect on food and drinking behavior, which can lead to food issues (Da Silva et al., 2002). The effect of insight on human physiology, including bone, muscle, hematology, fluid and electromagnet, is very important in various nutrition-related procedures (Iwase, Nishimura, Tanaka, & Mano, 2020). Nutritional criteria for long-term spatial travel have been identified on the basis of restricted studies and terrestrial extrapolation. In this respect the Nutritional Biochemistry Laboratory of NASA was responsible for developing space travel nutritional requirements (Smith SM; Lane HW, 2021). This is accomplished by both organizational and analysis programs. A nutritional status assessment program for all astronauts on the International Space Station. Biochemical and nutritional tests are conducted before, during, and after missions to meet this medical need. This program offers details on the crew's health and nutritional status, as well as efficacy assessments of countermeasures ("LSDA: Experiment - SMO 016E," 2014). Although flight physicians and scientists have confidence in the ability of the human body to recover from space flight and return to the environment of the earth, the results of the missions at Skylab, Mir, and NASA-Mir have persuaded flight doctors and scientists that actions and monitorings are essential to longer-term space flying performance. (Enrico, n.d.). Counter measurements are interventions that mitigate the harmful effects on humans in physical and psychological environments. It is necessary to maintain health and to permit countermeasures such as exercise, nutrition, and diet (Enrico, 2016). The key areas for food science and nutrition for future research are critical questions and a map (Tipton, 2015). This involves promoting and using research and development on genomics, proteomics, and other advanced technology (Norheim et al., 2012). The various disciplines need to collaborate in order to define the underlying processes and to use them to maintain a healthy and effective crew in terms of nutritional requirements. However, given the limited resources for spaceflight study, land modeling is essential for understanding physiological changes in the diet and its underlying mechanisms (Green & Johnson, 2015). Nutritional research requires a range of models and interdisciplinary methods, including contributions from the fields of physiology, biochemistry, psychology, food and technology, horticulture, and advanced medical technology (Sauberlich, 1984).

POSSIBLE SOLUTION FOR BREATHABLE AIR

Current air modernization systems are not suitable for potential human space programs including long-term space travel and inter-planetary missions due to their dependence on consumables such as water, stores and waste sorbents (Mulloth & Finn, 2021). The long-term tasks of transporting consumer products and supplies are costly for launch costs, and can jeopardize safety) (Drake, 2009). Future technologies can use and recycle consumables as much as possible to make long-term human space flights possible, and minimize the scale, length, and power demands. Due to the need for thermal energy both for CO_2 and for water disposal, CO_2 removal is one of the best.

A modern technology is in progress for alternative, less power CO_2 removal with molecules and new substrates. Current carbon TCCS beds are available to be expended. There are regenerative adsorbents available that can be used for controlling trace contaminants. Many processors operate separately from the existing air urban renewal scheme (Mulloth & Finn, 2021). Processor thermal integration will be applied in future systems. Current systems use an adsorption bed packaging configuration, which has a number of disadvantages, including a high-pressure drop and poor heat and mass transfer. For terrestrial usage, dark, thermally conductive substrate adsorbent immobilized have been developed and are commonly used (Rouhani, Huttema, & Bahrami, 2018).

GREATEST DANGER - RADIATION

The biggest danger posed by these risk factors is radiation. The most important aspect of space exploration is protection. Safety is paramount. It is evolving in response to the growing difficulty of space travel and the steady progression of human missions to conduct manned missions to the Moon, Mars, and beyond ("5 Hazards of Human Spaceflight," 2011). On this perilous course, we will encounter a slew of issues, including a slew of unknowns and natural- source threats in outer space, all of which will coexist with great technologies and breakthroughs. Among the natural issues, we consider cosmic radiation as the most dangerous factor of space exploration (Horneck et al., 2003). Outside the Earth's magnetosphere, solar and galactic cosmic rays (SCR and GCR) pose major threats to spaceflight protection and safety. Solar flares are particularly dangerous ("NASA - Understanding Space Radiation," 2021). Solar flares are highly complicated phenomena that detect a broad variety of energetic particles across the entire electromagnet spectrum, from radio wavelengths to extreme ultraviolet, X-rays, and gamma rays. Flares are classified according to their strength, duration, morphology, or magnetic topology, as well as their characteristic corpuscular radiation (Benz, 2008).

The behavior of the Sun shown in flares creates a strong flow of energetic particles interacting with the home planet and greatly disrupting our spatial atmosphere, known as space weather (Batu, 2019). The spacecraft operating and damage to onboard electronics may break down energy particles (Martinez & Kingston, 2012). They pose a great risk to astronauts, especially during spacewalks and off-road activities. Aerobic-free, air-rail-marine transportations, pipelines, geographical information system/data management (GIS, internet), just to name just a few, are affected by geomagnetical storms caused by solar flares and social-economicinfrastructure (kickinger, 2020). All of the components are now inextricably related and dependent on global space systems such as space communication, navigation, geodesy, oceanography, natural resource management, ecology, and so on. Any failure or disturbance in these global system functions has a significant impact on modern society, making it more vulnerable to solar activity (Martinez & Kingston, 2012). If systems grow more complex over time, the social and economic effects of space weather are more likely to increase. To - the painful effects of solar flares, varioussafety precautions must be taken.

The well-being of future astronauts is also a concern, given what we now know about spatial radiation and its effects on the human body. More research into the effects of high-LET radiation is required because it provides knowledge that can be used to improve mitigation strategies. Furthermore, better plans for securing satellites and future outposts on celestial bodies,as well as more research into radioprotective medicines, would be beneficial. Finally, by adopting the new IOM Decision System, NASA will be given the ethical support it needs to expose astronauts to higher levels of potential risk than was previously acceptable.

CONCLUSION

Risk Management is a key subject of future spaceflight flights. The support properties of structures transporting men to and from Mars are partially determined by decisions taken in the near future and are direct drivers of the costs and risks of lifecycle systems. This paper highlighted the many implications of cost support, performance, schedules and risks, and a number of approaches to address these impacts.

New support approaches are required to enable safe and successful spaceflight beyond LEO.

In the context of long-term spaceflight, there are substantial technical and physiological concerns related to diet. Nutrition is a vital part of maintaining the functionality of the endocrine and immune systems, the skeletal and muscle health, and the hydration status of space crews, all of which are necessary for long-term missions. In addition, food creates a moral team and promotes the productivity of psychosocial aspects. The full role of diet in the execution of long-term tasks is essential to this achievement. A blend of fundamental and applied research is required to ensure the timely delivery of results to decision-makers at both the strategic and organizational stages. Lastly, the exposure of people to risk demands the highest degree of ethical accountability, such that explorations are carried out without compromising the individual nature of the explorers.

In the end, new science and technology will make for a more autonomous, advanced therapeutic capability for low-earth missions. These advancements would not only mitigate the risk of a surgical evacuation of the low-earth astronauts but would also play a crucial role in preparing the next phase of human space exploration—missions beyond the Earth's immediate orbit. Without space medicine doctors, technological advancements, and human space exploration cannot make optimum and certain progress in ensuring the exploration of the civilization of explorers is carried out without compromise. Clinics specializing in the delivery of health care in a partial and microgravitational environment will be required for human space missions of increasing time and duration. Based on their experience in space medicine, these doctors will continue to work on an evidence plan to provide a clinical basis to help address human space flight medical challenges. Mastering these challenges will help free humans from physical gravitational limitations and lead an exploration into space beyond the earth.

References

5 Hazards of Human Spaceflight. (2011). Retrieved March 8, 2021, from NASA website:https://www.nasa.gov/hrp/5-hazards-of-human-spaceflight

Benz, A. O. (2008). Flare Observations. Living Reviews in Solar Physics, 5. https://doi.org/10.12942/lrsp-2008-1

Chambial, S., Dwivedi, S., Shukla, K. K., John, P. J., & Sharma, P. (2013). Vitamin C in DiseasePrevention and Cure: An Overview. Indian Journal of Clinical Biochemistry, 28(4), 314–328. https://doi.org/10.1007/s12291-013-0375-3

Chambial, S., Dwivedi, S., Shukla, K. K., John, P. J., & Sharma, P. (2013). Vitamin C in DiseasePrevention and Cure: An Overview. Indian Journal of Clinical Biochemistry, 28(4), 314–328. https://doi.org/10.1007/s12291-013-0375-3

Crewed Missions. (2021). Retrieved March 8, 2021, from Collinsaerospace.com website:

https://www.collinsaerospace.com/what-we-do/Space/Crewed-Missions

Cucinotta, F. A., & Durante, M. (2009). Risk of radiation carcinogenesis. Human health andperformance risks of space exploration missions. NASA SP-2009-3405. Houston: National Aeronautics and Space Administration, 119-170.

Da Silva, M. S., Zimmerman, P. M., Meguid, M. M., Nandi, J., Ohinata, K., Xu, Y., … Inui, A.(2002). Anorexia in space and possible etiologies. Nutrition, 18(10), 805–813. https://doi.org/10.1016/s0899-9007(02)00915-2

Douglas, G. L., Zwart, S. R., & Smith, S. M. (2020). Space Food for Thought: Challenges andConsiderations for Food and Nutrition on Exploration Missions. The Journal of

Nutrition, 150(9), 2242–2244. https://doi.org/10.1093/jn/nxaa188

Drake, B. G. (2009). Human Exploration of Mars Design Reference Architecture 5.0 Mars Architecture Steering Group NASA Headquarters ACKNOWLEDGEMENTS. Retrieved from
https://www.nasa.gov/pdf/373665main_NASA-SP-2009-566.pdf

Durante, M. (2014). Space radiation protection: Destination Mars. Life Sciences in SpaceResearch, 1, 2–9.
https://doi.org/10.1016/j.lssr.2014.01.002

Enrico, C. (n.d.). Space nutrition: the key role of nutrition in human space flight. Retrieved fromwebsite:
https://arxiv.org/ftp/arxiv/papers/1610/1610.00703.pdf

Furukawa, S., Nagamatsu, A., Nenoi, M., Fujimori, A., Kakinuma, S., Katsube, T., ... Hidema, J.(2020). Space Radiation Biology for "Living in Space." BioMed Research International, 2020, 1–25.
https://doi.org/10.1155/2020/4703286

Horneck, G., Facius, R., Reitz, G., Rettberg, P., Baumstark-Khan, C., & Gerzer, R. (2003). Critical issues in connection with human missions to Mars: Protection of and from the martianenvironment. Advances in Space Research, 31(1), 87–95. https://doi.org/10.1016/s0273-1177(02)00662-2

Jones, P. M. (2010). Human Performance in Space. Reviews of Human Factors andErgonomics, 6(1), 172–197.
https://doi.org/10.1518/155723410x12849346788787

Kim, H. W., & Rhee, M. S. (2020). Space food and bacterial infections: Realities of the risk androle of science. Trends in Food Science & Technology, 106, 275–287. https://doi.org/10.1016/j.tifs.2020.10.023

Langston, S. M. (2016). Space Travel: Risk, Ethics, and Governance in Commercial HumanSpaceflight. New Space, 4(2), 83–97.
https://doi.org/10.1089/space.2015.0015

Loades, M. E., Chatburn, E., Higson-Sweeney, N., Reynolds, S., Shafran, R., Brigden, A., ... Crawley, E. (2020). Rapid Systematic Review: The Impact of Social Isolation and Loneliness onthe Mental Health of Children and Adolescents in the Context of COVID-19. Journal of theAmerican Academy

of Child & Adolescent Psychiatry, 59(11), 1218-1239.e3.
https://doi.org/10.1016/j.jaac.2020.05.009

Manzey, D. (2004). Human missions to Mars: new psychological challenges and research issues. Acta Astronautica, 55(3-9), 781–790.
https://doi.org/10.1016/j.actaastro.2004.05.013

Marov, M. Ya. (2020). Radiation and space flight safety: An insight. Acta Astronautica, 176,580–590.
https://doi.org/10.1016/j.actaastro.2020.03.022

Martinez, L. M., & Kingston, J. (2012). Space radiation analysis: Radiation effects and particle interaction outside the Earth's magnetosphere using GRAS and GEANT4. Acta Astronautica, 72,156–164.
https://doi.org/10.1016/j.actaastro.2011.09.001

Mulloth, L. M., & Finn, J. E. (2021). Air Quality Systems for Related Enclosed Spaces:Spacecraft Air. The Handbook of Environmental Chemistry, 383–404. https://doi.org/10.1007/b107253

Mulloth, L. M., & Finn, J. E. (2021). Air Quality Systems for Related Enclosed Spaces:Spacecraft Air. The Handbook of Environmental Chemistry, 383–404. https://doi.org/10.1007/b107253

NASA - Understanding Space Radiation. (2021). Retrieved March 8, 2021, from Nasa.gov website:
https://www.nasa.gov/audience/foreducators/postsecondary/features/F_Understanding_Space_Radiation.html

Norbury, J. W., Schimmerling, W., Slaba, T. C., Azzam, E. I., Badavi, F. F., Baiocco, G., … Goodhead, D. T. (2016). Galactic cosmic ray simulation at the NASA Space Radiation Laboratory. Life Sciences in Space Research, 8, 38–51. https://doi.org/10.1016/j.lssr.2016.02.001

Norheim, F., Gjelstad, I., Hjorth, M., Vinknes, K., Langleite, T., Holen, T., … Drevon, C. (2012).Molecular Nutrition Research—The Modern Way Of Performing Nutritional Science. Nutrients, 4(12), 1898–1944.
https://doi.org/10.3390/nu4121898

Radiation Effects on Astronautic Fertility in Space: Deep Space Policy. (2020). Retrieved March8, 2021, from The Journal of Science Policy & Governance website:

https://www.sciencepolicyjournal.org/article_1038126_jspg_16_01_01.html

Rouhani, M., Huttema, W., & Bahrami, M. (2018). Effective thermal conductivity of packed bedadsorbers: Part 1 – Experimental study. International Journal of Heat and Mass Transfer, 123, 1204–1211. https://doi.org/10.1016/j.ijheatmasstransfer.2018.01.142

Sauberlich, H. E. (1984). Implications of nutritional status on human biochemistry, physiology,and health. Clinical Biochemistry, 17(2), 132–142. https://doi.org/10.1016/s0009-9120(84)90344-8

Sihver, L., & Mortazavi, S. (2019). Radiation Risks and Countermeasures for Humans on DeepSpace Missions. 2019 IEEE Aerospace Conference. https://doi.org/10.1109/aero.2019.8742175

Sudip Bhattrai, A., Magar, A., Kumar, R., Bhandari, G., & Adhikari, S. (n.d.). AIAA STUDENT DESIGN COMPETITION HUMAN SPACEFLIGHT: PHOBOS BASE. Retrieved from website:https://www.aiaa.org/docs/default-source/uploadedfiles/education-and-careers/university-students/design-competitions/tribhuvanuniversity-3rdplace.pdf?sfvrsn=f923d98f_0

Chapter 7. Long Duration Experiment Technical Report: Mars Advanced Detector In-Situ-Utilizer

Abstract

This report proposes the MADIE (Mars Advanced Detector In-Situ Utilizer Experiment), a long-duration experiment conducted on the surface of Mars by astronauts to explore the interior of Mars and the amount and analyst the liquid water present on Mars by redesigning the Mars Advanced Radar for Subsurface and Ionosphere Sounding (MARSIS), a European interplanetary mission to search for ice and water in the Martian surface. A group composed of astronauts and crews of one hundred members will go to Mars. The redesign of the MARSIS mission will allowastronauts to investigate the provenance and sustainability of liquid water under current Martian conditions. This paper provides an overview of MADIE's operation and modeling. Along with that, MADIE utilizes solid oxide electrolysis to transform carbon dioxide into oxygen to lower the cost of sending humans to Mars by using the atmosphere of Mars as its resource, nearing one step closer to Civilization on Mars. MADIE will be the first instrument of its kind to orbit the Earth, and its creation will be equally unique. As new data is added to the dynamic model, it becomes a quick and low-cost way to test MADIE without putting expensive equipment in a dangerous situation.

Introduction

Mars is a cold, dusty, and barren planet with a thin CO2-based atmosphere. However, the geomorphology and contrapuntal record of the surface shows that Mars has a thicker atmosphere and saltwater flowing on its surface in the past. (Sharp, 2017) As a result, it has been speculated that life might have evolved and that certain primitive life types may still survive today. This prospect has sparked a surge in interest in Mars, with all of the world's main space agencies already involved in its discovery. The first European interplanetary spacecraft, Mars Express, was planned to include an unparalleled wide coverage of the planet's surface, subsurface, and atmosphere, with an emphasis on two related matters(Orosei et al., 2015), the actual availability of ice or liquid water in the Martian crust, and potential signs of past or present bioactive compounds on the planet (Chicarro et al., 2004)

MADIE analyzes Mars' geophysics, with a focus on interior strategies and early evolution, provides critical constraints for models of Mars' thermal, geophysical, and geomorphologic evolution, as well as the proper application of SNC meteorite constraints.(Trotignon et al., 2001) Geophysical research can provide important constraints for Mars astrobiology by assisting in understanding why Mars lacks a global magnetic field, why Mars has undergone such drastic climate changes over its history, and by providing a limit for the intensity available to the chemoautotrophic hydrosphere through a measurement of the temperature stench. (Brromfiel, 2015)

MADIE applies the concepts of In Situ Resource Utilization. ISRU entailed the use of technology to turn commodities at a destination site into usable resources. The advancement of ISURU technology is a key element in the decrease of human spaceflight costs. (NASA MARS, 2020) The environment on Mars is 96% carbon dioxide, which can be transformed into oxygen by a process known as solid oxide electrolysis. (Hinterman & Hoffman, 2020) It is an ISRU method that is intended to minimize the cost of sending humans to Mars by using the planet's atmosphere as an advantage.

Additional Supplies

Since this is a long-duration flight, special equipment would be needed to support the expected two hundred total crew members after all of the rockets have been launched and arrived. To begin, eight rovers will be needed for each scientist to have the ability to use one for their study while on the Martian surface at some point. Also, two inflatable habitats would need to be designed and completely usable for the astronauts to remain in while being shielded from any potentially harmful radiation, regolith abrasion, or dust storm. (Mann, 2020) The Mars Advanced Detector In-Situ Utilizer Experiment digitalizes the incoming echoes of Mars and permanently places them on board for analog circuitry. This processing lowers the instrument's data rate to the maximum permitted by the spacecraft communications channel. (Trotignon et al., 2001) Astronauts on Mars evaluate the processed returns, which are also transmitted to Earth through the spacecraft's telecommunications system. Most of MADIE hardware is used for the ionosphere and subsurface sounders. The ionospheric sounding system provides a generic low pulse bandwidth that radiates through the antenna dipole. The sounder on the sub-surface works with a transmitting pulse of 250 μs over four potential frequency bands of 1 MHz bandwidth. A transmitter and receiver that share the 40 m tip to one end dipole antenna are included in the subsurface sonic channel. The dynamic range of the surface in places with a rough surface is strongly limited by surface clutter echos. The synthetic antenna processing reduces the off-nadir echoes from areas that stretch along the path.

In the Martian atmosphere, electronic radar instruments are used to track radar. In the electronics assembly, there are two packages. The digital electronics subsystems (DES) and receiver assembly are in the first package, while transmitters and antenna interfaces are in the third. All of these boxes are situated inside the Mars Express structure, except the antenna box, which is located on the spacecraft's exterior. The monopole antenna has a preamplifier on the structure's assembly layer. Each component has an antenna, which is made up of a fiberglass tube with an antenna conductor wire within. For the launch, the components are folded and mounted separately in a rectangular carrier. After landing on Mars, the components were launched one by one using a pyrotechnic system at right angles to the spacecraft solar array. (Chang Díaz et al., 2019) The 7-m monopole is designed similarly to the dipole elements, but with a smaller diameter and deployed at a right angle to the dipole, parallel to Mars' Z-axis.

MADIE, unlike MARSIS, transmits a modulated waveform after ground processing to achieve a range resolution equal to the transmitted bandwidth's inverse (Jordan et al., 2009). Range compression is a processing technique that entails matching the received echo to the waveform that was transmitted. (Trotignon et al., 2001) This gives the best resolution and significantly increases the signal-to-noise ratio (SNR), but it also produces side-lobes, which can mask secondary reflections that occur during surface reflection unless the signal is filtered further. Radargrams, or grey-scale images in which the horizontal dimension corresponds to distance along the ground track, the vertical dimension is the round trip time of the echo, and the brightness of the pixel is the logarithm of echo amplitude, are widely used to display data processed in this manner.

The SOXE subsystem, which uses water to produce oxygen from carbon dioxide, is also a component of MADIE. SOXE is the leading concept that drives oxidation reaction in MADIE. (Hinterman & Hoffman, 2020) An electrolyte membrane consisting of yttria-stabilized Zirconia (YSZ) is sandwiched between a cathode and an anode in a SOXE cell. The cathode is made up of a nickel sheet that sits on top of the YSZ and does not initiate properly. Since nearly all electrons are ionically exchanged with an oxygen ion, and four electrons are transferred per molecule of O2, the current is directly proportional to the oxygen production amount. (Hinterman & Hoffman, 2020)

Power Requirements

A small city requires an enormous amount of energy, which is the same situation for the Mars colony. Solar panels will be stored in central areas and used to fuel battery storage systems that will power fixed installations and recharge mobile craft batteries. (Chang Díaz et al., 2019) Solar panel deployment on Mars' surface is regarded as having very high longevity, medium strength, and very high-cost levels. The mission's risk decreases significantly as the scale of the solar array increases. Another benefit of lander-mounted solar power is that it quickly makes the lander self-sufficient for keep-alive power, removing the time constraint of autonomously connecting cables to a remote power farm until the on-board energy supply is depleted, resulting in medium capacity and extremely high costs. The "petals" of a solar range are folded, stretched, and screened to a deployment pad. (Chang Díaz et al., 2019) Composite bipers are rolling and spring-loaded to protect a thin film panel against deployment plates and petal stacks. The external boom deployment is passively powered by potential energy stored in the bi-stable booms which enable the device to be fully deployed.

Mission Timeline

Based on what is found in the carbon dioxide and water molecules of Mars in this mission, the astronauts will spend a total of four years collecting raw echos, electron density and studying frequency bands attenuated in the spectrum, and storing useful results in specially constructed containers. The sample will be held in a jar that will imitate the Martian atmosphere for long periods when living bacteria or animals will be detected. (Rincon, 2021) After four years, all of the crew members will be free to return to Earth to discover or resume their experiments with whichever sample they brought back with them on the spacecraft. When the samples are returned to the Earth's surface, a team of scientists with backgrounds in Physiology, Chemistry, and Geology will closely examine them to keep an eye on any live organisms while making a scientific conclusion on every specimen that was returned.

Mission Prediction

Scientists on Mars are expected to begin using the MADIE subsurface radar monopole antenna about 3 months into the experiment. They would have discovered differing frequencies produced by plasma of different densities at various altitudes, as well as liquid water under the atmosphere. After around six months of work, the scientists found primitive forms of life in the molten water flowing on the planet's surface. They would have penetrated about 15% of the surface and found a previously unknown mineral in the regolith, such as Magnesium or Strontium. After one year of pure operation to the Martian surface, scientists may have experimental data of conversion into oxygen through the mechanism known as solid oxide electrolysis and found one to two samples that they claim to be clear proof of life once on Mars by a kind of fossil or remains of a bacteria or organism that got left behind after it became expelled. (Rincon, 2021) They will have covered approximately 15% of the surface and will continue to work until they have analyzed the entire surface and obtained approximately twenty to thirty samples weighing approximately 0.10 oz each.

MADIE will be the first radar sounder technology to be in the mars, despite the many risks and difficulties it will face in the initial phase of the Mars Exploration Mission. The ionospheric soundings obtained by MADIE from the operation will make an important contribution to our knowledge of the Martian ionosphere. MADIE will produce unique data that gives proof of many questions about life forms on Mars. (Rincon, 2021) A few comparisons must be made, such as the differences between experimenting on Mars and Earth, the extra instruments required for the longer period, the power supplies to be used, and the approximate duration of the experiment, as in any experiment. Knowing whether there is or was life on Mars will help us keep our lives in perspective by proving that we are not alone in this vast and dark universe.

Conclusion

MADIE will be the first radar sounder technology to be in the mars, in spite of the many risks and difficulties it will face in the initial phase of the Mars Exploration Mission. The ionospheric soundings obtained by MADIE from the operation will make an important contribution to our knowledge of the Martian ionosphere. MADIE will produce unique data that gives proof of many questions about life forms on Mars. A few comparisons must be made, such as the differences between conducting an experiment on Mars and Earth, the extra instruments required for the longer span of time, the power supplies to be used, and the approximate duration of the experiment, as in any experiment. Knowing whether there is or was life on Mars will help us keep our lives in perspective by proving that we are not alone in this vast and dark universe.

References

Orosei, R., Jordan, R. L., Morgan, D. D., Cartacci, M., Cicchetti, A., Duru, F., ... Picardi, G. (2015). Mars Advanced Radar for Subsurface and Ionospheric Sounding (MARSIS) after nine years of operation: A summary. Planetary and Space Science, 112, 98–114.https://doi.org/10.1016/j.pss.2014.07.010

Trotignon, J. G., Séran, H.-C., Béghin, C., Meyer–Vernet, N., Manning, R., Grard, R., & Laakso,H. (2001). In situ observations of the ionized environment of Mars: the antenna impedance measurements experiment, AIM, proposed as part of the Mars advanced radar for subsurface and ionospheric sounding, MARSIS. Planetary and Space Science, 49(2), 155–164.https://doi.org/10.1016/s0032-0633(00)00134-3

Hinterman, E., & Hoffman, J. A. (2020). Simulating oxygen production on Mars for the MarsOxygen In-Situ Resource Utilization Experiment. Acta

Astronautica, 170, 678–685.https://doi.org/10.1016/j.actaastro.2020.02.043

Chang Díaz, F., Carr, J., Johnson, L., Johnson, W., Genta, G., & Maffione, P. F. (2019). Solarelectric propulsion for human mars missions. Acta Astronautica, 160, 183–194.

https://doi.org/10.1016/j.actaastro.2019.04.039

Researchers Reveal How Climate Change Killed Mars. (2015, November 5). Retrieved March22, 2021, from NPR.org website: https://www.npr.org/sections/thetwo-way/2015/11/05/454594559/researchers-reveal-how-climate-change-killed-mars

mars.nasa.gov. (2020). Mars Oxygen In-Situ Resource Utilization Experiment (MOXIE).Retrieved March 22, 2021, from Nasa.gov website:https://mars.nasa.gov/mars2020/spacecraft/instruments/moxie/

Moon. (2020, September 25). Moon safe for long-term human exploration, first surface radiationmeasurements show. Retrieved March 22, 2021, from Science | AAAS website:

https://www.sciencemag.org/news/2020/09/moon-safe-long-term-human-exploration-first-surface-radiation-measurements-show

Rincon, P. (2021, February 19). Nasa Mars rover: Key questions about Perseverance. RetrievedMarch 22, 2021, from BBC News website: https://www.bbc.com/news/science-environment-53129281

Chapter 8. Mars Manned Outpost Mission Proposal

Need: Understand the history and presence of life on Mars with the assistance of human crew on-site to provide more in-depth observational analysis than the unmanned rovers, while also establishing an atmosphere like Earth to investigate the possible prospects of life on Mars with the hope that possibility of human extinction could decrease by the colonization of other planets.

Goal: To find and experiment areas on the planet that have evidence of habitable environments, peculiarly those that suggest the permanence of liquid water in the past or present, to better support life in the future for colonization.

Objective: To dig deep into Mars' crust and look for fossils of living organisms such as primitive single-celled organisms and determine the geological process that has shaped the local terrain, iron-containing minerals, classify and quantify contingent amounts of distinct mineral types that comprise water or were created in water that can hint about life below the Mars' surface. Also, the mission strives to create a mutuality of benefits for all stakeholders through the mission's operations and seeks by putting the mission's Principles into actions every day to make a difference to people and the planet through extraordinary performance.

Mission: Safely transport Rovocopter with human crews in the quest for ancient life by Utilizing the data accumulated from previous un-crewed robotic missions to establish the first permanent inhabited settlement on Mars that aims to help develop the first permanent inhabited settlement on Mars to address many fundamental scientific questions regarding the origin and evolution of our solar system as well as the question of the possibility of life on Mars.

Introduction

Mars colonization will soon become the impetus for a new path of human evolution in space. Mars is an enthralling world that has always piqued scientists' interest, which is why so many robotic probes have been sent there. Mars provides valuable scientific expertise, persuasive stimuli for technological development, future great economic returns, biological and cultural stability, and the ability to enhance foreign affairs and a profound opportunity to advance human culture and consciousness in the cosmos, alongside exciting adventures. (Martian Outpost, 2013) Although a veritable fleet of robotic missions has reached Mars and returned with a wealth of scientific evidence, many scientific questions about Mars and its past remain unanswered. (Taraba et al., 2006).

It is assumed that Mars will provide the Earth's colonists with a far more hospitableatmosphere than any other nearby planet. Promising findings from rovers and a low-frequency radar mounted on the spacecraft in Mars Express have also long held the conviction that underground and sub-glacial water can be found (Orosei et al., 2018). In addition, like Earth, Mars, with the latest proof of metal ores and other important mineral substances, is predicted to have significant mineral resources on and under its surface. (Professional Brasser & Mojzsis, 2017). These questions will remain unanswered until humans set foot on Mars' surface and begin conducting scientific experiments beyond any robot's capabilities, no matter how well built. Building upon a robotic mission, a human crewwill deeply deepen the history of Mars and beyond to decide if life ever existed on Mars. Their works are preceded. They will also look for evidence of existence in the present and serve as the start of exploration into the future of Martian life by creating a colony to sustain it by influencing it chemically, by forming a terraform and being a viable habitat for human beings and other life forms. This mission aims to respond to the most important questions about Mars existence by looking at the evolution of Mars, its present situation, and the future possibilities. If there is life beyond our Earth, Mars manned exploration might address the deep query. (Slakey & Spudis, 2008) Therefore, research on the Red Planet is one of the most significant obstacles of this new century, which the mission aims to answer.

Mission Name

The name of this mission is MARSOC, which stands for Mars Advanced Rovocopter Surface Outpost Colonization Mission. Advanced Rovocopter means that technology integrated into this mission is very sophisticatedly integrated into Rovocopter. Rovocopter is a robotic device in which human will explore mars. The mission is performed through the robotic assets along with the astronauts. The Surface Outpost Colonization in the name of this mission means that with human and rovocopter, humans will gradually overcome the technological challenges of a deep space mission, the possibility of exploration and colonization of extraterrestrial outposts through the historical MARSOC mission.

Primary Goals

The MARSOC mission has numerous goals. The MARSOC mission primary goal is to create a permanent base colony on Mars from which high-valuable scientific and engineering research can be carried out in three main areas mentioned below:

On Mars, searching for life past and present.

Basic research in astronomy to learn new information on the nature and evolution of the solar system

Study in applied science about how Mars tools should be used to expand the life sustainability of structures and how to survive elsewhere

A predecessor to the MARSOC project, the robot mission explored the history, current, and future of Martian life. Today, Mars is explored by people in orbit and on the surface through robotic assets. Mars Base Camp is a way of accelerating scientific exploration of origins and the quest for life through scientific/astronauts and sophisticated robots in the Martian environment, by identifying the right areas for human landing and returning correct samples to previous field research. Mars Base Camp provides the experimental module with 7 metric tons of scientific equipment and 40 kW of electrical energy. (*Pathways to Exploration*, 2014) MARSOC mission's scientists would analyze samples by digging in the Mars crusts from the rover themselves, eliminating the necessity to send an image to the Earth. While digging for samples of fossils, the rover will discover many unique materials that will be used to build a Martian base for the MARSOC crews, which serves as another goal of this mission. MARSOC will further develop on the rover's data and continue the quest for current life on Mars by analyzing the samples.

The allocations aim to provide a starting point for discussions on the scientific objectives of the measuring types, the instruments and support equipment, sample cures, external robotic elements, interfaces, operational concepts, and the identification of driver devices as well as their ability to access large numbers of raw data than those transferred to Earth by Mars. A more thorough review of the Mars Base Camp mission model and the use of robotics/automated and human/human capabilities and structures to advance core scientific goals for the research community will be undertaken. (Mu ller, 2020) These resources are meant to include a basis for considerations of the scientific goals, type of measuring, experiment instruments, external robot components, protocols and practical and pe-conducting principles, as well as the ability to view and retrieve greater quantities of raw data relative to existing Mars ties to Earth. An in-depth study of the Mars Base Campdesign, with science group focusing on the most adequate activities, robotics, manuals, and structures, will be undertaken to further the core scientific objectives. The theory of robotic systems, modules and operations further than the Mars Base Camp is particularly important and accentuates the capabilities and deficiencies of the crew and storage arrays at the Mars Base Camp best. ("Schedule Management Handbook," 2011) NASA's current and potential aspirations, external partners' objectives and perhaps market priorities with a view to capital are all part of the Mars Base Camp concept. MARSOC will carry out the robotic mission's future feature to help Mars maintain life. This is the first opportunity for people to navigate the environment and provide helpful knowledge about the Earth's systems safely and effectively.

Date	Mission Event Description
May 2024	Study interior structures of Mars through collection of datafrom previous rovers
November 2025	Launch robot to Mars to perform Mars surface explorationwith sample cache
April 2026	Start collecting data from the Robotic mission and observed the data.
June 2027	Begin the assembly in Martian Space to provide an improved communication relay system that is essential forproviding high bandwidth communications back to Earth and images of Mars sites at high resolution.
April 2028	Start analyzing all the data gathered from the RoboticMission.
November 2029	Conduct Cislunar scientific exploration demonstrating tele- operations and sample retrieval capabilities
April 2030	Begin assembly at the Deep Space Gateway of Mars-boundMars Base Camp elements
December 2030	Approximate end of data analysis and synthesis from therobotic mission

January 2031	MARSOC mission planning in full force
April 2031	Begin assembly at the Deep Space Gateway of Mars-boundMars Base Camp elements
June 2032	Pre-deploy Mars Base Camp science assets with solarelectric propulsion
August 2033	End of construction of base
January 2034	Launch of MARSOC crew to Mars
June 2034	Arrival of MARSOC crew to Mars
March 2036	Conduct full system tests of the assembled Mars Base Campahead of departure for Mars accordingly as assigned in MARSOC objectives.
April 2038	Launch MARSOC crews back to Earth for depart
July 2040	Arrival of Crews in Earth

("Mars Base Camp: An Architecture for Sending Humans to Mars," 2017) The prospect of life on Mars will be studied from a settlement on a terraformed land recently.

MARSOC set goals based on a previous robotic mission to study life on Mars and support a colony of human investigators who will discover the earth in a manner that cannot be technologically advanced.

Mission Duration and Timeline

The MARSOC mission must last this long because the terraforming stage takes a long time to complete in order to send the crew. The MARSAC mission will begin shortly after the robotic mission concludes, allowing the earth to be quickly assembled. It will take the teamfive years to establish a successful foundation, conduct numerous tests, and observe terraforming processes in order to learn more about life on Mars.

Location

Site selection and reactions to the site are the major problems with the first settlement. Many technical and logistical aspects must be taken into account. The settlement must be adjacent to a geologically rich field of science. In-situ services, of which water is the most important, must be accessible easily. Low elevation is important to benefit from higher atmospheric pressure that benefits both gas collection and parachuting. (*Evolutionary and Revolutionary Technologies for Mining*, 2002) The location also provides convenient access to and from the orbit, as well as a clear view of the orbit and Earth communication system. According to some sources, a condition in the northern hemisphere is advantageous since the summer is long. [Dubbink 2001], [Robinson 1993].

The site should also be modularly designed. A variety of urban and architectural aspects are also present. The first habitat to be permanent is our first cut into western borders. It is therefore a crucial phase that will influence the earth's future development. (Almusaed & Almssad, 2020) There would simply be insufficient time to adopt a predetermined plan that ignores local conditions. The other side of a completely unexpected and tactical solution, such as that adopted by North American colonists, is unreal too. In a series of guidelines, the base must respond to the site. (Petrov, Mackenzie, Homnick, & Palaia, 2005) The platform also needs to be protected in a real and perceived way. On Earth, security considerations have historically dominated the location of a new settlement.

Since there are no indigenous peoples to defend against on Mars, a genuine psychological sense of security from the hostile environment will be more necessary. Each settlement must have a sign with which the people may identify, as well as a sense of belonging as a home or a holy site commemorating the group's founding. The first town's location would eventually become a place of pilgrimage and veneration. It is worth remembering.

To fulfill the requirements MASROC mission will be land it's rovocopter in Candor Chasma in order to dig in the Martian crusts for fossils and life in the present-day. This location has been chosen because Candor Chasma fulfills many of the requirement for a successful site. Candor Chasma is a northern subset of Valles Marineris. Since the canyon's floor is almost 4800 meters below the global mean, air pressure is relatively high. The walls of the canyon climb about 8 kilometers above the base of the canyon. (Petrov, Mackenzie,

Homnick, & Palaia, 2005)A magnificent panorama of mountains and valleys would surround the first area. In addition, the tables themselves are interesting and provide a narrow view of the settlement. Areas where the flight routes to and from Orbit will not flyover the city are located toward the north and south of the mesas. The picture of Candor Chasma shown below shows the lower background image quality superimposing all the high-resolution images available on the Mars Global Surveyor from the Mars Orbital Camera.

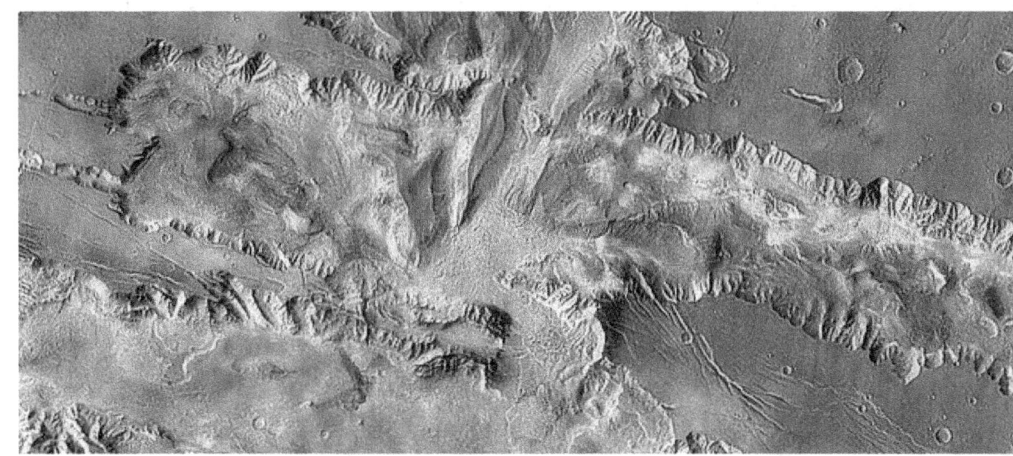

Candor Chasma, with portions of Ophir and Melas chasmata at top and bottom left, respectively, of Mars' massive Valles Marineris canyon system.

Mission Elements

The project components involve the necessary machinery and technology to carryout the task. A multi-stage rocket and rover payload were the mission components of the robotic mission that acted as precursors for the MARSOC mission, Rovocopter, which was used as the transport for rover and then as a contact system for rovers. A vision and a future roadmap are essential to each settlement. Throughout literature, there are several precedents. The Law of the Indians, the laws developed in Spain for the design of all colonial cities constructed in South American, is an example of influence for this initiative.

Another square grid that was offset by 45 degrees from the north was placed in every colonial town. The municipal buildings were facing the main square. The aristocracy remained on plots directly adjacent to the center, while most of the population inhabited the outlying squares. It was so successful that the organization began constructing housing towns in Chile's northern deserts in the early twentieth century. Though the basic features and rigidity of Indian law are irrelevant to the Martian world, MASROC will attempt to construct a vision of the future settlement that offers the same rigorously organized logic governing the relationships between various elements.

The NASA Reference Project is well-known for its complexity. The NASA study specifically notes that there is a methodological challenge because of the multiple dangers that must be considered and reduced to acceptable standards. (Salotti & Heidmann, 2014)The entrance, descent, and arrival stages, as well as the lengthy, complicated construction of massive low-Earth spacecraft and the use of low-tech thermal engines, are all major threats. Several studies indicate that other options allow for substantial risk reduction, despite the methodological concern. The main idea is to reduce the overall challenge by making a lot of simplifications.

The critical elements required for this MARSOC mission are:

1. Crew transportation beyond low-Earth orbit

2. Autonomous crew operations

3. Procedures of deep space staging

4. Ascension to Mars

5. Life support and habitation devices 6. Space radiation protection/shielding

6. Systems for entry, descent, and landing

7. Energy conservation and surface control

8. Surface mobility

9. Human robotic incorporation

10. In-situ resource utilization on Mars

11. Long-term human health

12. Techniques for deep space operations

13. Aerocapture of Mars

14. Atmospheric Earth entry at high velocity 17. Human factors, such as living inpartial gravity

Mission and System Requirements

The system components of the various mission elements listed above are included in this job. Alternatively, each task function may also involve a computer or subsystem dependingon the environment.

Device certification is very stringent in human space travel for safety reasons. From design and development to incorporation into the space vehicle, a long list of protocols and tests must be followed to obtain human-rated certification. The systems must be "flight- proven" for the same configuration and length as the human mission at the end of the operation. Many of the mission's and systems' specifications are enumerated below. These conditions must be met for the mission to be successful. (Salotti & Heidmann, 2014)

The propulsion systems must be evaluated for the same payload, vehicle configuration, and environmental conditions as those required for a human mission.

Without any assistance from other spacecraft, the habitable modules must be tested with the same life support equipment, the same number of astronauts, and the samelifetime requirements.

Aerocapture and EDL maneuvers must be practiced using the same vehicle shape and scale, as well as the same equipment and procedures.

The deployment and operation of in situ propellant production (ISPP) systems must be evaluated in the same setting and for the same amount of time as a human mission.

The other three fundamental requirements for the building of a Martian settlement are defense, reliability, and extensibility. Each scheme must integrate many layers of redundancy for protection. A living ecosystem, for example, must be separated into several interconnected and individually strained parts. Two exits from either space should be available in the case of a fire, accidental pressure loss, or any other failure. (Petrov, Mackenzie, Homnick, & Palaia, 2005) Furthermore, the absence of one section must not cut off the settlement's working portions from one another. The vast distance from Earth, as well as the scarcity of labor and resources on Mars, determine efficiency. (Savage, 2018) Expandability necessitates that the development pattern is conveniently repeatable and expandable without jeopardizing the quality of the existing frameworks. Before going into depth on the materials and building techniques to be used, it's a good idea to go through the technical problems that must be addressed.

Crew Responsibilities

For the MARSAC mission, a crew of twenty-five people would travel to Mars. They would have an unparalleled vision of the world and give the project greater autonomy than robotic missions since humans have decision-making and science skills that rovers lack.

The team will be responsible for the task's progress. (mars.nasa.gov, 2019) They'll take measurements including air composition and temperature readings to monitor the progress of terraforming. They will conduct research and assist in the hunt for life on Mars today. They'll also examine fossil samples and help in the quest for past life on Mars. (Crane, 2020)

There will be specialties of the crew selected from different expertise for the success of the MARSAC mission. The expertise will be ranging from builders, architects, farmers, and scientists. The construction firms will concentrate on habitat expansion. Their work will mainly take them outside of a small indoor studio's building or planning habitat.

Engineers oversee the life support system's construction and maintenance. They would develop new resources in the foundation's immediate vicinity. Mechanics serve more settlements than anybody else and have robust additional vehicles (EVA). (Starzee, 2018) They will have a hangar and restoration shop where rovers and external chemical plants will be able to be brought in for repair in a pressurized climate. They will be like the 'alchemists' from Stanley Robinson's Red Mars [Robinson 1993]. Farmers will primarily be employed in the plant preparation sector, which will aid in the production of plant-rated greenhouses. They will share some of the scientists' test rooms. The scientists will do fundamental research in the settlement's las and on long roving journeys.

There will always be at least one long roving mission. During the visits, specimens would be studied in an open laboratory and results would be synthesized in private areas or private neighborhoods. Finally, at the grassroots, a commander will guide and coordinate the work. Preparing and cleaning the common areas, for example, are separated or rotated equally. To establish an initial habitat, most settlers are construction workers, engineers, and fishermen, with only a limited number of researchers. More scientists will be included as the settlement matures and the population grows.

Risk Mitigation

For a manned mission to Mars, there is a lot to consider and schedule. In either case, precautions should be taken to reduce the risk. Radiation, team selection, meteorite collisions, crew health, and landing are only a few of the more frequent challenges. The outpost on Mars will have a variety of surface structures for supporting human life. Each elemental structure can perform a specific role that is vital to the outpost's survival. (Tran, 2019) Risk Mitigation for MARSOC depicts a typical architecture of the Mars outpost, highlighting main structures and possible device positions.

The example outpost represents a research and development milestone in the life cycle of the Mars surface system development. Power generation and delivery to power-dependent surface hardware are provided by central power systems, which are usually photovoltaic arrays and regenerative fuel cells. (Chancellor et al., 2018)

Human systems assist the crew directly in terms of the pressurized volumes of housing, labs, and airlocks as well as logistics containment rather than support mission tasks or other surface hardware. The most difficult task in the history of the outpost remains direct crew support. Human system capabilities include crew size accommodation, surface mission life, solar flare protection, long system life, automated failure detection, and recovery. ("VOSS Model: Mars," 2014) Of all output capacities, mission duration has the largest impact on human systems. The presence of an outpost indicates that the facilities are being used for everyday human visits. As a result, the machines can be used continuously or intermittently. Trips to the Martian outpost will take 8 years based on orbital opportunities. The duration of a surface mission has a major impact on the life support system (US) configuration. The decision is if the LSS's inputs and outputs have been secured, and if so if mechanical or biological locking is used. The quantity and volume of rasp ply needed to produce non-renewable waste products are referred to as closure. A fully closed LSS does not require resupply. (Mason, 2016) The two most important system parameters, weight, and power are affected by the various system configurations.

An outpost on Mars needs more than just secure infrastructure to keep a human presence going. Each surface structure, including scientific instruments, will include its own set of operating procedures. Surface activities include anything from automating atmospheric plants to assembling, checking, and inspecting housing systems. Creation, mission management, and support activities are the three operating categories that cover the wide range of outpost tasks. Each community has its own set of issues. (Harris, 2018)The initial outpost building activities would rely on the site and mating of the habitat and additional pressurized buildings. Habitat unloading, positioning, anchoring, and potential soil covering could all be done using automated or telerobotic techniques. Medium and long-haul operating service, task readiness plans, and crew protection and reliability maximizations will be used for mission operations. The material support policy and preparation strategy between the world and Mars can be emphasized by supporting operations. ("Asteroid Research Begins Under the Sea," 2011) They help challenge will also decide the correct protocol for shutdown and restart. The ultimate tactical challenge will be determining the optimal combination between Earth-based and Mars-based funding.

MARSOC demonstrates a cooperative approach to outpost activities on Mars. The outpost's operational obligation varies as time passes. (Williams, 2016) The outpost's initial purpose is to provide Earth-based soil assistance to Mars as well as crew surveillance and control systems. The surface operations have become more autonomous as the front end has changed to a well-established platform. Advances in Martian machinery applications and robotics technologies have made it possible to travel from Earth to Mars. (Shiga, 2020) Ground support is no longer needed as a result of the change to automated results. The mission's success is also aided by early support for the outpost. The process includes scenario descriptions, crew preparation, project planning, and software design. It is essential to develop. If the outpost is discovered and everyday operations are familiar, development tasks decrease.

Mars has a low magnetic field and a thin atmosphere that protects against cosmic radiation only sparingly. Three sources of radiation must be addressed. First, a flood of filled particulates is the solar wind that continually releases from our star. The other is cosmic rays produced by explosive stars in the galaxies that bombard the solar system from everywhere. The use of martian regolith in all habitability spaces has been found to protect the inhabitants of Mars against the first two sources of radiation (Zubrin 1996), (Wilson 1997). Water tanks or construction material containing a high proportion of hydrogen atoms (such as most polymers) frequently can shield them (Kim et. al. 1998). The final cause of hazard is solar flares, which occur as the sun expels massive clouds of charged particles. Wide solar flare radiation emissions necessitate a much higher level of safety. Fortunately, several spacecraft are observing the Sun and can alert Mars up to an hour ahead of time. As a result, the village must have a solar storm shelter that can hold all inhabitants during a solar flare.

Mission Constraints

The constraints of the mission are described in mission limits. NASA Headquarters set limits before 2040. The time for training for a launch, the selection of crews, and the Mars terraform are restricted. (Williams, 2019) Even, by 2045 at NASA headquarters, the crew must be returned to Earth. Another limiting factor for the MARSAC mission is the energy on the earth and Mars along with the inadequate capacity to provide Mars weight.

This promotes resourcefulness, including printing 3D on the base from already existing materials on Mars and choosing what is needed to keep the mass, and therefore the cost, to a zero. MARSAC is often restricted by the pace in progress of technology. The terrain deformation factor is particularly constrained by technology, because it's never performed before, and the Martian atmosphere must be shaped by physicists. (Achenbach, 2021) Like any job, time, money, and energy are extremely important as limitations.

Since a mission involves all of the restricting variables, foreign counterparts expand the scope of technologies acquired or developed. Besides, more countries and scientists have to provide inventions, science, jobs, access to technologies, and finance. The shore is a multinational mission that introduces fewer restrictions. (Howell, 2018) ESA will participate since much of the technologies integrated into MARSAC is from this organization though as the mission moves from a concept stage to a deployment level, other nations will be included.

References

Taraba, M., Zwintz, K., Bombardelli, C., Lasue, J., Rogler, P., Ruelle, V., ... Wessels, L. (2006).Project —a study for a manned Mars mission in 2031. *Acta Astronautica*, *58*(2), 88–104.
https://doi.org/10.1016/j.actaastro.2005.04.013

Martín-Torres, F. J., Zorzano, M.-P., Valentín-Serrano, P., Harri, A.-M., Genzer, M.,Kemppinen, O., ... McConnochie, T. (2015). Transient liquid water and water activityat Gale crater on Mars. *Nature Geoscience*, *8*(5), 357–361. https://doi.org/10.1038/ngeo2412

Do, S., Owens, A., Ho, K., Schreiner, S., & de Weck, O. (2016). An independent assessment ofthe technical feasibility of the Mars One mission plan – Updated analysis. *Acta Astronautica*, *120*, 192–228.
https://doi.org/10.1016/j.actaastro.2015.11.025

Brasser, R., & Mojzsis, S. J. (2017). A colossal impact enriched Mars' mantle with noblemetals. *Geophysical Research Letters*, *44*(12), 5978–5985.
https://doi.org/10.1002/2017gl074002

Orosei, R., Lauro, S. E., Pettinelli, E., Cicchetti, A., Coradini, M., Cosciotti, B., ... Noschese, R.(2018). Radar evidence of subglacial liquid water on Mars. *Science*, eaar7268. https://doi.org/10.1126/science.aar7268

Petrov, G. I., Mackenzie, B., Homnick, M., & Palaia, J. (2005). A Permanent Settlement on Mars: The Architecture of the Mars Homestead Project. *SAE Technical Paper Series*.https://doi.org/10.4271/2005-01-2853

Salotti, J.-M., & Heidmann, R. (2014). Roadmap to a human Mars mission. *Acta Astronautica*, *104*(2), 558–564.
https://doi.org/10.1016/j.actaastro.2014.06.038

Mars Base Camp: An Architecture for Sending Humans to Mars. (2017). Retrieved April 12,2021, from New Space website:
https://www.liebertpub.com/doi/full/10.1089/space.2017.0037

Slakey, F., & Spudis, P. D. (2008). Robots vs. Humans: Who Should Explore Space? *ScientificAmerican Sp*, *18*(1), 26–33.
https://doi.org/10.1038/scientificamerican0208-26sp

Müller, V. C. (2020). Ethics of Artificial Intelligence and Robotics (Stanford Encyclopedia ofPhilosophy). Retrieved April 12, 2021, from Stanford.edu website: https://plato.stanford.edu/entries/ethics-ai/

Almusaed, A., & Almssad, A. (2020). City Phenomenon between Urban Structure andComposition. *Sustainability in Urban Planning and Design*.
https://doi.org/10.5772/intechopen.90443

Schedule Management Handbook. (2011). Retrieved April 12, 2021, from NASA website:https://www.nasa.gov/content/schedule-management-handbook

Home, sweet martian home. (2018). Retrieved April 12, 2021, from Chemical & EngineeringNews website:
https://cen.acs.org/articles/96/i1/build-settlements-Mars-ll-need.html

mars.nasa.gov. (2019). Mars Exploration Rovers | Missions – NASA's Mars ExplorationProgram. Retrieved April 12, 2021, from NASA's Mars Exploration Program website: https://mars.nasa.gov/mars-exploration/missions/mars-exploration-rovers/

Crane, L. (2020). Terraforming Mars might be impossible due to a lack of carbon dioxide.Retrieved April 12, 2021, from New Scientist website:

https://www.newscientist.com/article/2175414-terraforming-mars-might-be-impossible- due-to-a-lack-of-carbon-dioxide/

Starzee, B. (2018, August 21). Clients expect special expertise from architects, engineers and builders. Retrieved April 12, 2021, from Long Island Business News website: https://libn.com/2018/08/21/clients-expect-special-expertise-from-architects-engineers- and-builders/

How NASA Will Protect Astronauts From Space Radiation at the Moon. (2019). Retrieved April 12, 2021, from NASA website: https://www.nasa.gov/feature/goddard/2019/how- nasa-protects-astronauts-from-space-radiation-at-moon-mars-solar-cosmic-rays

Chancellor, J. C., Blue, R. S., Cengel, K. A., Auñón-Chancellor, S. M., Rubins, K. H., Katzgraber,

H. G., & Kennedy, A. R. (2018). Limitations in predicting the space radiation health risk for exploration astronauts. *Npj Microgravity*, *4*(1). https://doi.org/10.1038/s41526-018-0043-2

VOSS Model: Mars. (2014). Retrieved April 12, 2021, from Purdue.edu website:https://engineering.purdue.edu/vossmod/mars.php

Mason, B. (2016, May 13). Help Map the First Human Outpost on Mars. Retrieved April 12,2021, from Science website: https://www.nationalgeographic.com/science/article/help- map-the-first-human-outpost-on-mars

(2018). Retrieved April 12, 2021, from Nss.org website: https://space.nss.org/settlement/nasa/spaceresvol4/influence.html

Asteroid Research Begins Under the Sea. (2011). Retrieved April 12, 2021, from NASAwebsite: https://www.nasa.gov/mission_pages/NEEMO/NEEMO15/undersea_asteroid.html

Shiga, D. (2020). Space radiation hits record high. Retrieved April 12, 2021, from New Scientist website: https://www.newscientist.com/article/dn17882-space-radiation-hits-record-high/

Williams, M. (2016, November 21). How bad is the radiation on Mars? Retrieved April 12,2021, from Phys.org website: https://phys.org/news/2016-11-bad-mars.html

Williams, M. S. (2019, June). How easy or difficult would it really be to colonize our neighbouring planet, Mars? This articles looks at all. Retrieved April 12, 2021, from Interestingengineering.com website:

https://interestingengineering.com/making-a-garden-on-the-red-planet-how-could-we-colonize-terraform-mars

Howell, E. (2018, May 11). What Are Cosmic Rays? Retrieved April 12, 2021, fromSpace.com website: https://www.space.com/32644-cosmic-rays.html

ABOUT THE AUTHOR

The journey all begins in the land of the Himalayas, Nepal. I've had a keen interest in many things from my infancy. My childhood was full of memories, excitement, and entertainment. In Nepal, I researched and learned about the machines and how they functioned when the technology was implemented in my community. I was intrigued by the way computers operate. Internet was just introduced in Nepal however I did not know how to use the internet or work that could be conducted on the web at that time. So, I only played games on the computer, and throughout my elementary school, I had concentrated more on how machines work overtime.

In my freshmen year of my high school , I saw a flyer promoting Space Science classes which involved engineering, science, and technology. After applying to this program, I surprisingly got accepted, then my thirst for Space Exploration expanded. I learned about each aspect of Space Exploration and gathered new ideas throughout the course, and it kept expanding my knowledge through continuous learning opportunities and professional projects. This made me inquisitive, inspired by scientists, engineers, and technologists from the Wallops Fight facility. I learned many things from which I aimed at changing people's lives for the better at experiencing new challenges: highs and lows and at continuing to learn, not for myself, but as an understanding for other people. I believe that is an essential part of bringing the people together, regardless of whether they be cooperation, initiative, or leadership. Although one thing was constant through all these things, it was a love of learning. It was hard to find new ways to express my astronomical physics while researching topics highly related to astrophysics. I applied for another program my junior year and got into the Virginia Aerospace program along with the summer academy. Both Virginia Space Coast Scholars and Virginia Aerospace Science and Technology Scholars shaped my curiosity toward Mars exploration, propelling me to combine the research I have done in both fields.

OTHER BOOKS BY BISWESH DHUNGANA

The Exploration and Conquest of the New World

www.ingramcontent.com/pod-product-compliance
Lightning Source LLC
Chambersburg PA
CBHW071521220526
45472CB00003B/1102